MESSENGER RNA RESEARCH PERSPECTIVES

MESSENGER RNA RESEARCH PERSPECTIVES

TAKEO TAKEYAMA
EDITOR

Nova Biomedical Books
New York

For permission to use material from this book please contact us:
Telephone 631-231-7269; Fax 631-231-8175
Web Site: http://www.novapublishers.com

LIBRARY OF CONGRESS CATALOGING-IN-PUBLICATION DATA

Messenger RNA research perspectives / Takeo Takeyama (editor).
p. ; cm.
Includes bibliographical references and index.
SBN-13: 978-1-60021-814-9 (hardcover)
ISBN-10: 1-60021-814-8 (hardcover)
1. Messenger RNA. I. Takeyama, Takeo. [DNLM: 1. RNA, Messenger. 2. Gene Expression. 3. Genetic Code. 4. Genetic Markers. QU 58.7 M584 2008]
QP623.5.M47M47 2008
572.8'8--dc22

2007030045

Published by Nova Science Publishers, Inc. ✦ New York

Contents

Preface

mRNA (messenger RNA) is the mediating template between DNA and proteins. The information from a particular gene is transferred from a strand of DNA by the construction of a complementary strand of RNA through a process known as transcription. Next three nucleotide segments of RNA, called tRNA (transfer RNA), which are attached to specific amino acids, match up with the template strand of mRNA to order the amino acids correctly. These amino acids are then bonded together to form a protein. This process, called translation occurs in the ribosome, which is composed of proteins and the third kind of RNA, rRNA (ribosomal RNA. This book presents new research in the field.

Short Communication A p38 mitogen-activated protein kinase is activated by pro-inflammatory stimuli and is generally recognised as a as a key signalling molecule whose activity is required for the expression of a range of pro-inflammatory mediators. For this reason considerable effort has been invested into the development of inhibitors of p38 and its downstream kinase, MK2, for the therapy of chronic inflammatory diseases such as rheumatoid arthritis.

Our understanding of p38 function has been greatly accelerated by the development of specific small molecule inhibitors of the kinase. These pyridinyl imidazole compounds were found to potently inhibit cytokine production induced by pro-inflammatory stimuli. The authors and others showed that the mRNAs of inflammatory mediators, such as COX-2 (Dean et al., 1999; Ridley et al., 1998), were destabilised following inhibition of p38. They concluded that p38 MAPK stabilises COX-2 mRNA and together with others went on to show that other pro-inflammatory transcripts are regulated in a similar fashion (Clark et al., 2003). All p38-regulated transcripts contain AU-rich elements in their 3'-untranslated regions (3'-UTR). These elements promote rapid deadenylation and rapid decay of the body of the mRNA. p38 stabilises mRNA by inhibiting ARE-mediated mRNA decay (Lasa et al., 2000; Winzen et al., 1999). p38 specifically inhibits deadenylation rather than blocking the decay of the mRNA body (Dean et al. 2003).

Short Communication B Objectives: Amelogenin, enamelin, sheathlin (ameloblastin/ amelin), enamelysin, and KLK4 are the major structural proteins and proteinases in developing tooth enamel. Recently, it has been shown odontoblasts express the amelogenin. In this study, the authors examined whether or not the other enamel proteins and proteinases were expressed in the rat odontoblasts.

Methods: The authors prepared secretory-stage enamel organ epithelia (EOE), maturation-stage EOE, and odontoblasts lining dentin under the secretory and maturation stages enamel from the labial site of the developing incisors. Total RNA of each cell layer sample was obtained with the Total RNA Miniprep kit (Stratagene, USA). Single-strand cDNA was prepared from 3μg of each total RNA sample using a Ready to go T-primed First-strand beads kit (Pharmacia, USA). Each of the cDNA was used for PCR. In addition, rat incisor was fixed by Karnovsky's solution. After dehydration, the specimens were embedded in Technovit 7200VLC (Kulzer, Germany) and cut by Exact cutting machine (Exact, Germany). The sections were polished until 50μm and double stained with Stevenel's blue and Van Gieson picro-fuchsin.

Results: Histological examination was revealed that the highly mineralized enamel existed as a narrow layer along the enamel-dentin junction. The mRNAs encoding the structural enamel proteins and proteinases were found more or less in odontoblast samples. The amelogenin, enamelin, and sheathlin mRNAs were expressed odontoblasts under the secretory stage enamel. The enamelysin and KLK4 mRNAs were expressed both EOE and odontoblasts. The enamelysin mRNA was found in the secretory EOE. For KLK4, this trend was reversed. The KLK4 mRNA expression in maturation EOE was higher in secretory EOE. Furthermore, the KLK4 mRNA was expressed more in odontoblasts than in the secretory EOE.

Conclusion: These results indicate that the rat odontoblasts express enamel proteins and proteinases mRNA, which may play some part in the formation of dentin.

Chapter I - Diverse structural regularities identified in the genetic code, over the last four decades, have been unified with a model of code evolution that equated the time-order of amino acid (aa) addition to the code with the number of reaction steps (from the citrate cycle) required for its synthesis. To help establish the mechanism that coordinated growth of aa synthesis pathways with code formation, evidence was sought that tRNA adaptor molecules functioned as cofactors in aa synthesis, prior to appearance of the last common ancestor (LCA).

Phylogenetic analysis of the conserved trace of ancestral tRNA revealed: (1) LCA tRNA species that descended from a common ancestor contained similar anticodons and were adaptors for aa synthesized from a common precursor. (2) LCA tRNA adaptors for sibling aa generally possessed the same core structure group and had elevated sequence identity. (3) Five domains of contiguous codons read by tRNA, sharing the same core group and bearing sibling aa, were spread along code rows. (4) These coding domains covered most of the code. They also overlapped extensively with code regions assigned to same-family aa and with regions read by phylogenetically-related LCA tRNA. (5) Type ID tRNA were adaptors for five aa, including three NH_4^+ fixers, synthesized on 1-2 step paths. A tRNA-ID was found to be ancestral to the type IA tRNA asparagine-adaptor, which, in turn, displayed kinship with adaptors for fourteen longer-path (4-14 steps) aa. (6) Nearly all extant intermediates in aa synthesis, in accord with the foregoing results, were noted to contain a free tRNA attachment site.

Based on these observations, tRNA-dependent aa synthesis was credited with producing code domains by channeling nearest-neighbor codons, in code rows, to successive generations of synthetically related aa, during initial column-by-column code expansion.

Putative cofactor/adaptor molecules normally appeared to diversify from a sibling aa tRNA; principally, an amide-amino-acid-adaptor. Primordial aa precursors, aspartate and glutamate, were identified as likely substrates for the first aminoacyl-tRNA synthetase enzymes. The pattern of aa specificity among class I and II synthetases was attributed to the imprint of code structure on synthetase evolution.

Chapter II - RT-PCR targeting amoA mRNA which coding a subunit of ammonia monooxygenase (AMO) was combined with fingerprinting methods to elucidate effect of some important environmental conditions on transcriptional activity of amoA gene. Expression of amoA gene in domestic wastewater treatment process was monitored by combination analysis of RT-PCR and DGGE. Sludge samples were collected from bench-scale reactors. After total RNA was extracted, RT-PCR targeting amoA mRNA was performed. DGGE analysis was used to separate the amplified products according to their sequence and thereby to visualize individual community members. Ammonia oxidizers that were demonstrated by the presence of amoA mRNA were regarded as metabolically active ones. The DGGE profiles of amoA mRNA RT-PCR products obtained from the samples collected at the initial stage of the reactor operation showed decrease of the diversity of amoA mRNA. When influent load was changed with reduction of hydraulic retention time (HRT) or increase of concentration of wastewater, new DGGE bands newly appeared on only the short HRT condition. The amoA sequences of DGGE bands affiliated to the Nitrosomonas eutropha lineage disappeared during the beginning of the operation. On the other hand, all the amoA sequences of new DGGE bands affiliated to the genus Nitrosomonas, and formed a phylogenetic lineage including no cultured representatives. Effect of ammonium concentration and temperature on transcriptional activity of amoA gene was also investigated. Activated sludge was collected from artificial domestic wastewater treatment reactor and incubated with different ammonium concentration (25, 100 and 250 mg l^{-1}). The incubation temperature was adjusted to 20 or 30°C. Ammonium consumption rate during the first few hours was higher at higher temperature and lower initial ammonium concentration. The T-RFLP profiles of amoA mRNA showed different effect of ammonium concentration and temperature on transcriptional activity of amoA for Genus Nitrosomonas and Nitrosospira. The observed main peak in the all samples was of the length associated with Nitrosomonas-like amoA (219 bp). Other peaks of Nitrosomonas-like amoA (491 bp) and Nitrosospira-like amoA gene (283 bp) were hardly detected in the samples incubated at 30°C. These minor peaks were observed in the samples incubated at 20°C and the height of these peaks were relatively high in the samples incubated with high initial ammonium concentration. Cloning analysis revealed that the dominant sequence of expressed amoA mRNA is not closely affiliated with that of any isolates. These results of mRNA analysis suggest that uncultured but active Nitrosomonas strains dominated in high ammonium concentration and high temperature condition.

Chapter III - In humans, wasting of lean body tissues occurs in many pathophysiological situations such as cancer, sepsis and trauma where starvation and undernutrition are also a frequent complicating factor. Many studies on protein metabolism have been performed in the whole body and skeletal muscles during acute wasting. Muscle atrophy during fasting is reflected by increased gene expression of protein breakdown components. Apart from muscles, several non-muscular tissues may also contribute to protein wasting in food-deprived situations. The interest in using long-term fasting as a model to study protein loss in vivo is that this state is notably marked by a successive low and high level of negative nitrogen balance. Long-term fasting is depicted by a brief period of adaptation (phase 1) which is followed by a period of large mobilization of lipids and of whole body protein sparing (phase 2) that ends with late enhanced protein utilization (phase 3). In this study, the authors investigated to what extent the gene expression of the main proteolytic system components found in various rat tissues was altered 1) during phase 2 and phase 3 characterizing prolonged fasting, 2) during refeeding and 3) in fasted animals treated with leptin. Fasting differentially up-regulated the expression of proteolytic-related genes in a more pronounced manner in glycolytic and mixed (tibialis, gastrocnemius) than in oxidative (soleus and heart) muscles and in other tissues. As expected, MAFbx (muscle atrophy F-box) and MuRF-1 (muscle ring finger 1) were up-regulated in the catabolic phase 3 of fasting, but also to a lesser extent during phase 2, notably in tibialis and diaphragm. Upon refeeding, the restoration of mRNA levels close to fed rat values was rapid (1 day) whatever the previous duration of fasting. Recombinant leptin administration resulted in no change of proteolytic-related gene expressions. In conclusion, the authors study underlines the tissue- and phase-specific heterogeneity in the regulation of proteolytic-related genes in the fasting/refeeding paradigm.

Chapter IV - One major difficulty in the diagnosis and treatment of renal cell carcinoma (RCC) is the limited sensitivity of techniques for the initial detection. RCC lacks efficient serum markers. The diagnosis depends solely on the imaging analysis. Recently, many studies have focused on the discoveries of molecular markers for tumors. Several markers of messenger RNAs (gene markers) have been available for RCC. The use of messenger RNAs to solve the clinical problems in RCC is relatively new and promising. In this chapter, the authors describe their personal experience on the study of the messenger RNAs in renal tumors and how they used these markers in clinical applications. The authors used the traditional method as well as developed a convenient assay on the messenger RNA of renal tumor samples. The messenger RNAs that the authors have studied are: MN/CA9, cadherin-6, vimentin, S100A1, etc. The authors further used the gene marker in cytoponction for the differential diagnosis, in the blood for detection of circulating renal cancer cells as well as in the biopsy for subtyping of renal tumors. All these studies demonstrate that the messenger RNAs are useful means to solve the clinical problems in the field of RCC.

Chapter V - The pre-mRNA polyadenylation in the nucleus is a critical event in eukaryotic gene expression since it is involved in messengers transport, stability and translation regulation. The enzyme responsible for the poly(A) tail addition is the canonical poly(A) polymerase that belongs to the DNA polymerases β-like nucleotidyltransferases superfamily. Recently, a distinct polyadenylation reaction has been described for certain transcripts involved in growth, cell cycle, development, genome maintenance, synaptic plasticity and long-term memory storage. This reaction is catalyzed by a different family of poly(A) polymerases identified as TRF4, GLD-2 and Cid proteins in Homo sapiens, Xenopus laevis, Caenorhabditis elegans, Saccharomyces cerevisiae and Schizosaccharomyces pombe.

In Entamoeba histolytica, the protozoan parasite responsible for human amoebiasis, the authors have previously reported the identification of the pre-mRNA 3′end cleavage and polyadenylation machinery. The authors have characterized the poly(A) polymerase EhPAP, which shares high sequence homology with canonical nuclear PAPs from other species. Here, the authors showed that recombinant EhPAP was able to bind RNA, even when it does not have a conserved RNA binding domain. Next, the authors extended their studies on mRNA polyadenylation and described a novel family of PAPs in this pathogen. By genomic survey and in silico analysis the authors identified six TRF4/GLD-2/Cid orthologous proteins in E. histolytica. Predicted proteins sequences are highly divergent from classical eukaryotic PAPs, but they possess the hallmarks of the DNA polymerases β-like nucleotidyltransferases superfamily, including the amino acids involved in nucleotide interaction and catalysis. E. histolytica non canonical PAPs have no apparent RNA recognition motif, which is critical for RNA binding. By RT-PCR assays the authors showed that genes are expressed in in vitro cultured trophozoites but with differential expression levels. Finally, phylogenetic and PSI-BLAST analysis showed that the PAP central domain of conventional and non conventional PAPs appears to have emerged from a bacterial ancestor known as the minimal nucleotidyltransferase (MNT) domain. The striking diversity of poly(A)-adding enzymes in E. histolytica suggests a greater degree of control over poly(A) tail length and mRNA stability than it has already been thought in this protozoan parasite. The further functional characterization of E. histolytica TRF4/GLD-2/Cid orthologous proteins would provide new insights into the biological roles of polyadenylation and mRNA turnover in this pathogen.

In: Messenger RNA Research Perspectives
Editor: Takeo Takeyama, pp. 1-8

ISBN: 978-1-60021-814-9

Expert Commentary A

Comments on the Search for the Source of the Genetic Code

_Brian K. Davis_*[*]
Research Foundation of Southern California, Inc.
P.O. Box 13595; La Jolla, CA, USA

After the standard code was established, in the mid-1960s, a search for its source began. Four decades later, it may reasonably be asked: Why has this endeavor taken so long? In the mid-1950s, after all, an essentially correct model for the axial structure of DNA was furnished concurrent with publication of the X-ray diffraction pattern for its B form and only three years after Chargaff's base ratio rules appeared. From this viewpoint, the question is: Why did the time-order of amino acid addition to the code remain unresolved for decades, when Franklin's photo B51, Chargaff's rules and some elementary nucleotide chemistry quickly led to the DNA double helix? [1] With sixty-four triplets assigned to twenty amino acids, it is true that the code possessed a more complex structure. Nonetheless, the correlation between amino acid path-distance and regularities in code structure [2] could have been discovered at any time after the code became known, since the amino acid synthesis pathways were well known. It is noticeable that early attempts to find the source of the code commonly relied on imported concepts of coding. With the benefit of hindsight, they appear to have frequently deflected attention from a fruitful avenue of research.

Structural symmetry between DNA strands encoding genetic information, not surprisingly, motivated attempts to find a stereochemical explanation for the code. It is noteworthy that shortly before the Watson-Crick model appeared, Dounce proposed, for non-stereochemical reasons, that nucleotides somehow encoded the amino acids incorporated into proteins. Then shortly post-Watson/Crick, the first of numerous stereochemical models was formulated by Gamow. [3] His model contained overlapping triplets that prohibited a number of residue pairs. The excluded pairs were soon found, however, in protein residue sequences.

[*] Brian K. Davis: email: davis@resfdnsca.org

So Gamow's 'diamond code' had been disproved, long before the genetic code itself was known.

When the code was revealed, Woese drew attention to a conspicuous cluster of codons for hydrophilic residues in the NAN column and another for hydrophobic residues in the NUN column. He credited these clusters to a difference in amino acid affinity for base triplets from each code column. Recent findings reveal that the clusters of polar and hydrophobic amino acids in the code correlate strongly with amino acid path-distance. [2] Fajan-Path's law indicates this pattern arose as the early code evolved toward a hydrophobic attractor in a mineral surface system. This finding, combined with freezing of the code, also negated theories ascribing these clusters wholly to selection forces directed at minimizing the deleterious effects of point mutations on protein function. These theories presupposed random substitutions extensively shuffled codon assignments of the standard set of amino acids, when, to the contrary, path-distance evidence indicates the tempo of change at advanced stages of code evolution approached zero.

By 1980, codon assignments were known to be mutable. A decade later, crystallographic studies began to reveal that an amino acid does not contact encoding nucleotides during activation, tRNA acylation and translation. These observations, together with the dependence of code structure on amino acid path-distance in ancient synthesis pathways [2], confirmed that the code relied on an indirect recognition mechanism. On finding elevated codon nucleotide levels in the putative binding site of RNA aptamers specific for Arg, Tyr, and Ile, Yarus credited a stereochemical mechanism with forming the genetic code, contrary to the foregoing findings. [4] Upon rescaling the nucleotide distributions he reported (Fig. 3 in [4]), in order to refer to triplets (coding entities in translation), anti-Ile- and anti-Tyr-aptamers showed no significant (NS) elevation in codon frequency. In addition, the significance (S) of Arg codons decreased by approximately four orders of magnitude: $\chi^2_{df=1}$(Tyr) = 9.48 [ref. 4] $\rightarrow$ 1.28 (codon probability = 0.13 NS, 1-tail distribution); χ^2_1(Ile) = 0.086 [4] $\rightarrow$ 0.107 (codon prob. = 0.37 NS); and, χ^2_1(Arg) = 26.6 [4] $\rightarrow$ 7.64 (codon prob. = $2.89\text{x}10^{-3}$ S). According to this, the proposed stereochemical coding mechanism rests on evidence from a single, basic amino acid.

The Arg result appears questionable, however. Chemical and physical methods were used to locate the binding site in each aptamer, with incongruous results. Three-dimensional NMR structure analysis revealed Yang's anti-Arg aptamer contained eight (8/33) nucleotides that interacted with the amino acid ligand. [5] A single Arg encoding triplet, 4-AGG, occurred among binding nucleotides; non-binding nucleotides interspersed throughout the binding site are disregarded, since they conceivably allow neutral substitutions. It is found, however, Arg codons are likely to arise simply by chance in nearly every fifth permutation of the Yang aptamer sequence;

$$p(x \geq 1) = 1 - (1 - 6.54\text{x}10^{-3})^{31} = 0.184 \text{ NS} \tag{1}$$

where the term with an exponent gives the probability that the aptamer contains no Arg codons, x = 0. The second term within parenthesis is the probability of forming an Arg codon by randomly sampling the eight binding nucleotides identified in NMR analysis and the exponent is the maximum number of base triplets in a 33-residue sequence. Yang's anti-Arg aptamer may be concluded to show no evidence of an elevation in Arg codons, among nucleotides revealed by NMR spectroscopy to bind the amino acid ligand.

A codon assignment specifies a qualitative relationship between base triplet and amino acid. Hence, the logic of using a quantitative parameter (frequency of binding-site codons) to connect an aptamer with formation of the genetic code also seems questionable. Similar objections are raised by subsequent studies that claim to implicate anticodons, as well as codons, in amino acid recognition by RNA aptamers. It remains unclear, moreover, how binding-site triplets from polyphyletic RNA molecules could 'leak' into anticodons of monophyletic tRNA species, during formation of all, or part, of the code. Development of the extant code, with its reliance on indirect recognition and diverse structural regularities linked to amino acid path-distance, in the Yarus scenario, requires unlearning an earlier direct-recognition coding mechanism. Although selection-amplification experiments have produced many RNA molecules with novel functions, they represent only a historic possibility. To link SELEX-produced RNA molecules to the origin of life, the conserved trace of RNA evolution, in the era preceding the last common ancestor, would require direct investigation.

Crick rejected early suggestions that a set of stereochemical interactions underlay coding specificity. This led directly to his prescient 'adaptor hypothesis'. On the other hand, his opposition to persistent support for a stereochemical basis to the code led, incorrectly, to the 'frozen accident' scenario of code formation. [6] A purely random code would not merit investigation, so interest in the code subsequently diminished among leading molecular biologists. It is now apparent that the code is far from random. Most interestingly, the path-distance model reveals the Standard Code has conserved vestiges of earlier codes. [2] The concept of efficient information encoding, imported from computer engineering, also proved to be unhelpful. [3] The genetic code turned out to be highly degenerate and, in a further display of nature's profligacy, junk DNA can far exceed gene sequences.

The antiquity of the genetic code accounts for the influence advances in origin-of-life research had on origin-of-code investigations. Foremost among them was Miller's demonstration, in 1953, that simple gases (methane, hydrogen, ammonia, water vapor) form small, thermally stable amino acids, and other substances, in silico, during prolonged exposure to electric discharge. [7] Occurrence of a Miller-type spectrum of small organic molecules in meteorites formed at the origin of the solar system, 4.6 billion years ago, furthermore, demonstrated abiotic organosynthesis had occurred in space. The impact of these advances is apparent from a survey of forty papers. [8] With each predicting a different temporal order of amino acid addition to the genetic code, nine of the first ten most nominated candidates occur in the abiotic soup produced by Miller. In an early theory linking code evolution and biosynthesis of new amino acids, it was considered necessary to assume that all codons were initially assigned to ten abiotically produced amino acids. [9]

tRNA adaptors for four amino acids produced abiotically (Miller synthesis) form a tetrad distinguished by complementary acceptor-stem second bases and complementary anticodons, whose mid-base (35N) matches (adopting wobble equivalence, G $\leftrightarrow$ A, C $\leftrightarrow$ U) the cis second base: tRNA-Gly (2**C**, 34G**C**C); -Ala, (2**G**, 34G**G**C); -Asp, (2**C**, 34G**U**C); -Val, (2**G**, 34G**A**C). [10] These amino acids were assumed to be primitive, given their possible abiotic origin. Furthermore, complementarity among the tetrad of tRNA adaptors purportedly conformed with an ancestral code being located in the acceptor-stem. Both orientations of the

G:C base pair at tRNA site 2:71 and two classes of ribozymal synthetases (r-aaRS), each cognate with a different strand of the acceptor-stem double helix, were implicated in coding these four amino acids. This '2x2' (GC) code evolved into the extant (GCAU) genetic code and, according to the theory, double-strand coding gave way to single-strand coding. The discovery that several synthetases specifically acylate a mini- or micro-helix segment of tRNA [11] motivated this model of code formation.

Showing that complementarity between acceptor-stem second bases in tRNA, with complementary anticodons, was a general feature of tRNA structure would clearly support the hypothetical ancestral acceptor-stem code. [12] Among twenty-five pairs of pre-divergence tRNA with complementary anticodons in Table 1, fourteen pairs contain complementary acceptor-stem second bases; a level of second base complementarity that is attributable to chance. Dual acceptor-stem/anticodon complementarity among tRNA pairs in the 3'-NGN/5'-NCN column (Table 1), likewise, conforms with a chance occurrence. Contrary to earlier reports [10, 12], these findings indicate that tRNA do not conserve evidence of an ancestral (G,C) acceptor-stem code.

Also central to an acceptor-stem origin of the code is catalysis of adaptor aminoacylation by two classes of aaRS, each cognate with a distinct acceptor-stem strand. tRNA phylogenetics and code structure, however, discount the proposed mechanism. They provide compelling evidence that most coded amino acids were synthesized homotopically, while attached to a tRNA cofactor, during formation of the code [13]. It follows that r-aaRS were solely responsible for charging tRNA with a precursor amino acid. In addition, this evidence shifts forward the time of origin of protein-synthetases (p-aaRS), for a specific amino acid, to late in the pre-divergence era. Displacement of tRNA cofactors then plausibly coincided with the protein takeover of amino acid synthesis pathways.

It was also suggested that the two-fold nature of synthetases and two-fold orientation of the G:C base pair at tRNA site 2:71 "presaged the core four column organization of the classic genetic code." [12] tRNA-Val and -Asp in the primary tetrad [10] are respectively acylated by class I and II p-aaRS. Amplifying this pattern to the code, it might be surmised that the columnwise distribution of class I and II p-aaRS in the code derives from the ancestral '2x2 code'. In particular, tRNA adaptors for amino acids encoded by NUN column triplets are acylated at the 2'-OH by class I p-aaRSs, and an anomalous class II Phe-RS. While amino acids encoded by NCN column triplets are attached to the 3'-OH of A76 by class II p-aaRSs. Contrary to the acceptor-stem model, however, a strong correlation exists between the columnwise pattern of codon assignments and amino acid path-distances. [2] Consequently, the columnwise aaRS pattern can be attributed instead to the imprint of code structure on the evolution of p-aaRS. This conforms with: (1) exclusion of r-aaRS for product amino acids by an apparent reliance on tRNA-dependent amino acid synthesis; (2) synthesis of p-aaRS by an advanced code in code-directed translation; (3) phylogenetic kinship among p-aaRS for NUN, and NCN, encoded amino acids, respectively [13]; and (4) a non-significant number of tRNA pairs (14/25) in Table 1 that pair class I and II p-aaRS (data not shown). Also notable is an acceptor-stem 2C:71G base pair in six (6/7) tRNA cognate with NAN column codons. NH_4^+ fixer amino acids, or amino acids with NH_4^+ fixer amino acid precursors, charge five of these tRNA. Significantly, these amino acids acquired tRNA that branch deeply in a phylogenetic tree for pre-divergence tRNA. [13]

Table 1. Frequency of complementary second bases in acceptor stem among pairs of amino acid adaptors with complementary anticodons

amino[2] acids	second base			dual compl.[3]		amino[2] acids	second base			dual compl.
	tRNA stem	anti-codon	compl.				tRNA stem	anti-codon	compl.	
3'-NUN/5'-NAN						3'-NGN/5'-NCN				
1. Asp/ Val	**C** C,**G**	CUG/ GAC	+	-		14. Ala/ Gly	**G** **C**	C**G**G/ G**C**C	+	+
2. Glu/ Phe	C C	CUU/ GAA	-	-		15. Ala/ Cys	**G** **C**,G	C**G**U/ G**C**A	+	+
3. Glu/ Leu	C C	CUC/ GAG	-	-		16. Ala/ Arg	G U	CGC/ GCG	-	-
4. Asn/ Val	**C** C,**G**	UUG/ GAC	+	-		17. Thr/ Gly	C C	UGG/ GCC	-	-
5. Lys/ Phe	N C	UUU/ GAA	-	-		18. Thr/ Ser	C N	UGA/ GCU	-	-
6. Lys/ Leu	C C	UUC/ GAG	-	-		19. Thr/ Cys	**C** C,**G**	UGU/ GCA	+	-
7. His/ Val	**C** **G**	GUG/ CAC	+	-		20. Thr/ Arg	**C** **G**,C	UGC/ ACG	+	-
8. Gln/ Leu	C C	GUU/ CAA	-	-		21. Pro/ Gly	**G** **C**	G**G**G/ C**C**C	+	+
9. Gln/ Leu	**G** **C**	GUC/ CAG	+	-		22. Pro/ Trp	G G	GGU/ CCA	-	-
10 Tyr/ Val	**C** **G**	AUG/ UAC	+	-		23. Pro/ Arg	**G** **C**,G	G**G**C/ C**C**G	+	+

Table 1. (Continued).

amino[2] acids	second base			dual compl.[3]		amino[2] acids	second base			dual compl.
	tRNA stem	anti-codon	compl.				tRNA stem	anti-codon	compl.	
11 STOP /Leu	C	/UAA				24. Ser/ Gly	C,G U	AGG/ UCC	-	-
12 Asp/ Ile	**C** **G**	CUG/ GAU	+	-		25. Ser/ STOP	G	AGU/		
13 His/ Met	**C** **G**	GUG/ CAU	+	-		26. Ser/ Arg	**G** **C**	**AGC**/ UCG	+	+
						27. Ala/ Ser	G N	CGG/ GCU	-	-
Total:			7						7	5
Prob.(second-base complementarity)[4]: $p(x \geq 14) = 0.346$ NS Prob. (dual complementarity, 3'-NGN/5'-NCN column): $p(x \geq 5) = 0.206$ NS										

[1] Based on consensus sequence for tRNA species in the last common ancestor, identified following analysis of 1100 sequences obtained in a search of the Bayreuth database (http://www.uni-bayreuth.de/departments/biochemie/tRNA/). Adapted from results of a previous investigation. [13]

[2] Pairs 4-6, 13, 17-19, 27 contain anticodons with a G:U wobble pair, as formation of pairs with a canonical 36A:34U base pair was hindered by the low frequency of anticodons containing a 5'-A (34-N) in pre-divergence tRNA sequences.

[3] dual compl., refers to tRNA species showing dual complementarity: acceptor-stem second bases (G,C) complementary in a tRNA pair, with complementary anticodons and with *cis* anticodon mid-base (35-N) and second base (2-N) identical. Bold letters highlight bases at both sites in these tRNA pairs.

[4] Second base and dual complementarity had non-significant binomial probability. Complementarity among G, C bases at site 2:71 had an intrinsic probability, $\nu = \frac{1}{2}$ per tRNA pair. Dual tRNA complementarity based on complementary G, C bases at site 2:71 and identical bases at *cis* site-2 and -35, had probability, $\nu = \frac{1}{2} \times \frac{1}{2} = \frac{1}{4}$ per tRNA pair in the 3'-NGN/5'-NCN column.

Soon after it became known, Crick suggested the code had evolved in three phases: [14] "1. The Primitive Code, in which a small number of amino-acids were coded by a small number of triplets. 2. The Intermediate Code, in which these primitive amino-acids took over most of the triplets in the code in order to reduce nonsense triplets to a minimum. The codons produced by this process for any one amino-acid were likely to have been related. ... 3. The Final Code, as we have it today." This intuitive scheme closely parallels the main phases of code evolution uncovered by analysis of amino acid path-distances: [2] Phase 1. NH_4^+-Fixers Code with four NH_4^+-fixing amino acids and a STOP signal encoded by sixteen base triplets. Phase 2. Expansion by addition of ten amino acids (hydroxy, S-containing, alkyl side chains) that formed domains of synthetically related amino acids, while acquiring the remaining forty-eight unassigned triplets. Phase 3. Overprinting by six basic and aromatic amino acids, through codon capture within error-prone boxes, to complete the Standard Code. Notwithstanding their similarity, three decades elapsed between Crick's scheme and the realization that structural regularities in the code correlate with amino acid path-distance. [2]

The source of this delay becomes apparent in his following paper, [6] where Crick states (p. 373), "Adenine is likely to be the commonest base available in the primitive soup ... the triplets containing only A's or G's in their first two bases (AA•, AG•, GA•, GG•) do indeed code for some of the more obviously primitive amino acids." This corner contains codons for Asp, Glu, Ser and Gly, which occur as racemic mixtures in preparations from abiotic sources. Crick clearly accepted the prevailing view in the mid-1960's: life began in a molecular soup (Miller's heterotrophic origin-of-life scenario). Not until prokaryote phylogenetics revealed autotrophic species aggregated close to the root of the tree of life [15], did a significant shift away from the heterotrophic theory take place.

The chemo-autotrophic origin-of-life theory coupled amino acid synthesis pathways to the autocatalytic, CO_2 fixing reductive citrate cycle. [16, 17] Analysis of these ancient pathways subsequently provided the 'road map' leading to a comprehensive theory of code formation, based on the real-time time-order of codons in precursor-product amino acid pairs. [2] Unaided by borrowed concepts, this theory incorporates a source for the code.

REFERENCES

[1] Klug A. (2004). The discovery of the DNA double helix. *J. Mol. Biol. 335,* 3-26.

[2] Davis BK. (2006). Making sense of the genetic code with the path-distance model. In: Ostrovskiy M. H., Editor. *Leading-Edge Messenger RNA Research Communications*. New York: Nova Science, 1-32.

[3] Hayes B. (1998). The invention of the genetic code. *Amer. Scientist. 86,* 8-14.

[4] Yarus M. (2000). RNA-ligand chemistry: A testable source for the genetic code. *RNA 6,* 475-484.

[5] Ellington AD; Khrapov M; Shaw CA (2000). The scene of a frozen accident. *RNA 6,* 485-498.

[6] Crick FHC. (1968) The origin of the genetic code. *J. Mol. Biol. 38,* 367-379.

[7] Miller SL; Orgel L. (1974) *The Origin of Life on the Earth.* Englewood-Cliffs: Prentice-Hall.

[8] Trifonov EN. (2000). Consensus temporal order of amino acids and evolution of the triplet code. *Gene. 261*, 139-151.

[9] Wong J T-F. (1975). A co-evolution theory of the genetic code. *Proc. Natl. Acad. Sci. USA. 72*, 1909-1912.

[10] Rodin S; Rodin A. (2006). Origin of the genetic code: first aminoacyl-tRNA synthetses could replace isofunctional ribozymes when only the second base of codons was established. *DNA Cell Biol. 25*, 365-375.

[11] Schimmel P; Giege R; Moras D; Yokoyama S. (1993). An operational RNA code for amino acids and possible relationship to genetic code. *Proc. Natl. Acad. Sci. USA. 90*, 8763-8768.

[12] Rodin SN; Ohno S. (1996). Four primordial modes of tRNA-synthetase recognition, determined by the (G,C) operational code. *Proc. Natl. Acad. Sci. USA. 94*, 5183-5188.

[13] Davis BK. (2007). Imprint of early tRNA diversification on the genetic code: domains of contiguous codons read by related adaptors for sibling amino acids. In: Takeyama T., Editor. *Messenger RNA Research Perspectives.* New York: Nova Science (at printer).

[14] Crick FHC. (1967). Origin of the genetic code. *Nature. 213*, 119.

[15] Woese CR; Kandler O; Wheelis ML. (1990). Towards a natural system of organisms: proposal for the domains Archaea, Bacteria and Eukarya. *Proc. Natl. Acad. Sci. USA. 94*, 5183-5188.

[16] Hartman H. (1975). Speculations on the origin and evolution of metabolism. *J. Mol. Evol. 4*, 359-370.

[17] Wachterhauser G. (1992). Groundworks for an evolutionary biochemistry: the iron-sulphur world. *Prog. Biophys. Mol. Biol. 58*, 85-201.

In: Messenger RNA Research Perspectives
Editor: Takeo Takeyama, pp. 9-16

ISBN: 978-1-60021-814-9

Expert Commentary B

Tristetraprolin Family Proteins May Prevent and Treat Cardiovascular Diseases

Heping Cao*

Diet, Genomics and Immunology Laboratory, Beltsville Human Nutrition Research Center, Agricultural Research Service, US Department of Agriculture, Building 307C, BARC-East, 10300 Baltimore Avenue, Beltsville, MD 20705, USA

Cardiovascular diseases (CVD) are the most deadly diseases in the US. According to the American Heart Association statistics [2], more than 64 million American men, women, and children suffer one or more types of CVD. CVD have been consistently ranked the No. 1 killer since 1900 (except 1918), accounted for 38.5% of all deaths in 2001, and was estimated to cost \$368.4 billion in 2004 in the US [2]. Congestive heart failure (CHF), one of the major cardiovascular diseases, resulted in 52,828 deaths and 995,000 hospital discharges in 2001 and was estimated to cost \$25.8 billion in 2004 in the US [2]. In this section, I reviewed recent information on the regulation of cytokine mRNA stability by tristetraprolin (TTP) family proteins in the cardiovascular system and discussed the potential applications of TTP family proteins in the prevention and treatment of CVD.

Cytokines Involved in Cardiovascular Diseases

The causes of CVD are complex. However, recent studies have pointed out the importance of pro-inflammatory mediators in the development and progression of CVD [8,38]. These mediators include tumor necrosis factor alpha (TNFα) and interleukin (IL) families [22,36,48,52]. Extensive studies have shown that TNFα is involved in numerous processes in the cardiovascular system, including affecting myocardial remodeling in

* Heping Cao: E-mail: *Heping.Cao@ars.usda.gov; peacetd2003@yahoo.com*

progression of heart failure [9], upregulating angiotensin II type 1 receptors on cardiac fibroblasts, a growth factor involved in post-myocardial infarction remodeling [28,44], reducing nitric-oxide synthase gene expression through post-transcriptional regulation of mRNA stability [1], and inhibiting endothelial nitric-oxide synthase gene promoter activity in bovine aortic endothelial cells [3]. According to the "cytokine hypothesis" [38], many aspects of the heart failure syndrome can be explained by the known effects of pro-inflammatory cytokines. Accumulation of high concentrations of these cytokines in the heart and the peripheral circulation could mimic some aspects of the heart failure phenotype, including left ventricular dysfunction, pulmonary edema, cardiomyopathy, endothelial dysfunction, anorexia, cachexia, receptor uncoupling from adenylate cyclase, activation of the fetal gene program, and cardiac myocyte apoptosis [38]. For these reasons, extensive efforts have been directed to the reduction of TNFα in heart failure patients by anti-TNF therapies using small molecules and by immunomodulatory strategies that activate anti-inflammatory pathways [9,31,53]. However, the clinical trials with TNF antagonists including etanercept and infkiximab have failed, probably due to their intrinsic toxicities and unanticipated effects in the setting of heart failure [38].

An alternative way to reduce inflammatory cytokine concentration in heart patients is to accelerate degradation of multiple mRNAs coding for the cytokines. It is well known that cytokine mRNAs are highly unstable following transcription in the nucleus and translocation in the cytosol. The instability of cytokine mRNAs is primarily due to the adenylate and uridylate-rich element (ARE) in the 3′-untranslated region (3′-UTR) [4], which is a major target for destabilizing proteins [39], and was proposed to be involved in cardiovascular pathophysiology [40].

Tristetraprolin Family Proteins Involved in Cytokine mRNA Instability

Tristetraprolin (TTP) and related proteins are well-established factors which destabilize cytokine mRNAs [6,21]. TTP is the best-understood member of a small family of tandem CCCH zinc finger proteins. Similar zinc finger sequences are found in at least 19 species in the GenBank database ranging from human to yeast and plants (including human, cow, mouse, rat, sheep, chimpanzee, dog, horse, pig, Xenopus, carp, zebrafish, Drosophila, C. elegans, baker's yeast, fission yeast, oyster, rice, and Arabidopsis.) In humans, three members of this family have been characterized: TTP (also known as ZFP36, TIS11, G0S24, and NUP475), ZFP36L1 (also known as TIS11B, cMG1, ERF1, BRF1, and Berg36), and ZFP36L2 (also known as TIS11D, ERF2, and BRF2) [6]. They are encoded by different genes and their patterns of cell- and tissue-specific expression and agonist-stimulated expression are quite different. However, they share certain properties: all have highly conserved tandem zinc finger domains, in which each C8xC5xC3xH zinc finger is preceded by the sequence R/KYKTEL, the two fingers are separated by 18 amino acids [6]; all are nuclear – cytoplasmic shuttling proteins [42,46]; and all are capable of binding AREs within single stranded RNA [7,10,12,20,33,34,54], and promoting the deadenylation and subsequent destruction of those transcripts, both in transfection studies and in cell-free experiments

[19,20,33,34]. In animals, TTP deficiency causes a profound inflammatory syndrome with cachexia, dermatitis, erosive arthritis, autoimmunity and myeloid hyperplasia, apparently due to excessive production of TNFα and GM-CSF, both of whose mRNAs are direct targets of TTP and are stabilized in cells from the knockout (KO) mice [19,33,51]. For these reasons, TTP can be thought of as an anti-inflammatory protein or arthritis suppressor [10,45,51].

TTP is an extraordinarily low-abundance, inducible, stable cytosolic, and hyperphosphorylated mRNA binding protein [10-12,17,18]. TTP mRNA and protein are detected in a number of tissues including spleen, thymus, lymph node, lung, liver, and intestine, consistent with its proposed anti-inflammatory role [17,35]. The expression levels of TTP mRNA and protein in mammalian cells are increased in response to several kinds of stimuli, including insulin and other growth factors, and stimulators of innate immunity, such as the endotoxin lipopolysaccharide (LPS) [17,35]. TTP is an mRNA-binding protein with high binding specificity for the so-called class II AREs within the 3′-untranslated mRNAs [4,7,10,12,19,20,32,33,54]. TTP and its related proteins can bind to the 3'-untranslated AREs of certain clinically important mRNAs, such as those coding for TNFα [10,12,20,33], GM-CSF [19], IL-3 [32], cyclooxygenase-2 [47,50], and plasminogen activator inhibitor type 2 [55]. This specific binding of TTP to the AREs results in the destabilization of the mRNAs in transfected human embryonic kidney (HEK) 293 cells [20,33].

Potential Involvement of TTP Family Proteins in the Cardiovascular System

Recent studies have provided evidence for the potential involvement of TTP and related proteins in the cardiovascular system. Two independent studies have shown that the expression of TTP in the heart is induced following thoracic transverse aortic constriction in wild-type mice [5,29] but not in knockout mice deficient in signal transducer and activator of transcription-6 (STAT6) [29]. Further results showed that TTP promoter is activated by endothelin-1 in the isolated neonatal cardiomyocyte from the wild-type but not the knockout mice [29]. STAT6 is a transcription factor that mediates many cytokine-induced responses and is activated by angiotensin II in rat hypertrophied hearts and in human hearts with dilated cardiomyopathy. The above results imply that TTP is involved in STAT6-mediated pathogenesis of cardiac hypertrophy and heart failure.

Vascular endothelial growth factor (VEGF), an angiogenic cytokine, has been proposed as the most important regulator of physiological and pathological angiogenesis [25]. The balance between pro- and anti-angiogenic factors regulates angiogenesis. VEGF increases vascular permeability and is mitogenic for endothelial cells. VEGF is involved in tumor progression, collateral vessel formation in ischemic tissues, inflammation, as well as in the development of diabetic retinopathy [25]. VEGF is also a key mediator of CVD [26,37] and adipogenesis in obesity [43]. Accumulation of VEGF above normal levels may lead to proliferation of leaky, friable vessels, a characteristic of diabetic eye disease. Uncontrolled neovascularisation within atheromatous plaques may lead to plaque growth and haemorrhage from these vessels contributing to plaque rupture [37].

One recent study has provided direct evidence for the involvement of TTP and related proteins, TIS11B and TIS11D, in the regulation of VEGF mRNA instability [23]. VEGF mRNA level is regulated by hypoxia, growth factors and hormones through both transcriptional and post-transcriptional mechanisms. VEGF mRNA has a short half-life and its abundance is regulated by the binding of stabilizing (HuR [27] and hRNP-L [49]) and destabilizing [23] proteins to its 3'-untranslated region. The ACTH-regulated TTP family proteins was shown to interact with the 3'-untranslated region of VEGF mRNA and decrease its stability (half-life reduced from 130 to 60 min) in vitro and in living cells through its AU-rich motifs [23]. Knocking down TIS11B expression in primary adrenocortical cells with small interfering (si)RNAs indicated that TIS11B participates in the control of both basal and, to a larger extent, ACTH-induced VEGF mRNA expression levels. TIS11B is the first protein identified so far that destabilizes VEGF mRNA and, therefore, may provide a new potential target in anti-angiogenic therapies.

Future Research on the Roles of TTP Family Proteins in Cardiovascular Diseases

Limited information is available about TTP in the cardiovascular system because of the difficulties in detecting TTP protein despite its readily detectable mRNA in the normal heart [17,35]. An alternative approach could be to address the causes preceding CVD, such as obesity and diabetes. Obesity is a major factor contributing to the development of CVD [30]. Some of the potential targets include TNFα and VEGF in macrophages and adipocytes of fat tissues. Both cytokine mRNA molecules are potential targets of TTP family proteins [20,23]. It was shown that plasma VEGF levels in obese mice and in athymic mice implanted with 3T3-L1 adipocytes in visceral fat are significantly higher than control and related to adiposity [41]. Plasma VEGF levels in patients with type 2 diabetes treated with insulin are 16% less than those treated with diet alone, and VEGF mRNA levels in 3T3-L1 adipocytes are increased by troglitazone and rosiglitazone, two antidiabetic TZD compounds shown to cause weight gain and edema [24].

Developing alternative strategies for the prevention of CVD also require further research. Our recently studies indicate that nutritional supplements such as cinnamon and green tea extract may help to reduce diabetes and obesity via TTP-mediated cytokine mRNA destabilization [13-16,23]. More detailed study is required to elucidate the molecular mechanisms underlining the regulation of cytokine mRNA stability by the TTP family proteins in the cardiovascular system and related tissues, thereby providing potential targets for developing alternative strategies for the prevention and treatment of CVD.

References

Alonso, J., Sanchez de Miguel, L., Monton, M., Casado, S. and Lopez-Farre, A. (1997). Endothelial cytosolic proteins bind to the 3' untranslated region of endothelial nitric oxide synthase mRNA: regulation by tumor necrosis factor alpha. *Mol. Cell. Biol.* 17, 5719-5726.

American Heart Association. (2004). *Heart Diseases and Stroke Statistics* - 2004 Update, American Heart Association, Dallas, Texas.

Anderson, H. D., Rahmutula, D. and Gardner, D. G. (2004). Tumor necrosis factor-alpha inhibits endothelial nitric-oxide synthase gene promoter activity in bovine aortic endothelial cells. *J. Biol. Chem.* 279, 963-969.

Bakheet, T., Williams, B. R. G. and Khabar, K. S. A. (2003). ARED 2.0: an update of AU-rich element mRNA database. *Nucl. Acids Res.* 31, 421-423.

Baumgarten, G., Knuefermann, P., Kalra, D., Gao, F., Taffet, G. E., Michael, L., Blackshear, P. J., Carballo, E., Sivasubramanian, N. and Mann, D. L. (2002). Load-dependent and -independent regulation of proinflammatory cytokine and cytokine receptor gene expression in the adult mammalian heart. *Circulation.* 105, 2192-2197.

Blackshear, P. J. (2002). Tristetraprolin and other CCCH tandem zinc-finger proteins in the regulation of mRNA turnover. *Biochem. Soc. Trans.* 30, 945-952.

Blackshear, P. J., Lai, W. S., Kennington, E. A., Brewer, G., Wilson, G. M., Guan, X. and Zhou, P. (2003). Characteristics of the interaction of a synthetic human tristetraprolin tandem zinc finger peptide with AU-rich element-containing RNA substrates. *J. Biol. Chem.* 278, 19947-19955.

Blum, A. and Miller, H. (2001). Pathophysiological role of cytokines in congestive heart failure. *Annu. Rev. Med.* 52, 15-27.

Bradham, W. S., Bozkurt, B., Gunasinghe, H., Mann, D. and Spinale, F. G. (2002). Tumor necrosis factor-alpha and myocardial remodeling in progression of heart failure: a current perspective. *Cardiovasc. Res.* 53, 822-830.

Cao, H. (2004). Expression, purification, and biochemical characterization of the antiinflammatory tristetraprolin: a zinc-dependent mRNA binding protein affected by posttranslational modifications. *Biochemistry.* 43, 13724-13738.

Cao, H., Deterding, L. J., Venable, J. D., Kennington, E. A., Yates, J. R., III, Tomer, K. B. and Blackshear, P. J. (2006). Identification of the anti-inflammatory protein tristetraprolin as a hyperphosphorylated protein by mass spectrometry and site-directed mutagenesis. *Biochem. J.* 394, 285-297.

Cao, H., Dzineku, F. and Blackshear, P. J. (2003). Expression and purification of recombinant tristetraprolin that can bind to tumor necrosis factor-alpha mRNA and serve as a substrate for mitogen-activated protein kinases. *Arch. Biochem. Biophys.* 412, 106-120.

Cao, H., Hininger-Favier, I., Kelly, M. A., Benaraba, R., Dawson, H. D., Coves, S., Roussel, A. M. and Anderson, R. A. (2007) Green tea polyphenol extract regulates the expression of genes involved in glucose uptake and insulin signaling in rats fed a high fructose diet. *J. Agri. Food Chem.* 55, 6372-6378.

Cao, H., Kelly, M. A., Kari, F., Dawson, H. D., Urban, J. F., Jr., Coves, S., Roussel, A. M. and Anderson, R. A. (2007) Green tea increases anti-inflammatory tristetraprolin and decreases pro-inflammatory tumor necrosis factor mRNA levels in rats. *J. Inflamm.* (Lond) 4,1, 12 pages.

Cao, H., Polansky, M. M. and Anderson, R. A. (2007) Cinnamon extract and polyphenols affect the expression of tristetraprolin, insulin receptor, and glucose transporter 4 in mouse 3T3-L1 adipocytes. *Arch. Biochem. Biophys*. 459, 214-222.

Cao, H., Polansky, M. M., Blackshear, P. J. and Anderson, R. A. (2006) Insulin and cinnamon polyphenols increase the amount of insulin receptor b, glucose transporter 4, and anti-inflammatory protein tristetraprolin in mouse 3T3-L1 adipocytes. *FASEB J.* 20, A93.

Cao, H., Tuttle, J. S. and Blackshear, P. J. (2004). Immunological characterization of tristetraprolin as a low abundance, inducible, stable cytosolic protein. *J. Biol. Chem.* 279, 21489-21499.

Carballo, E., Cao, H., Lai, W. S., Kennington, E. A., Campbell, D. and Blackshear, P. J. (2001). Decreased sensitivity of tristetraprolin-deficient cells to p38 inhibitors suggests the involvement of tristetraprolin in the p38 signaling pathway. *J. Biol. Chem.* 276, 42580-42587.

Carballo, E., Lai, W. S. and Blackshear, P. J. (1998). Feedback inhibition of macrophage tumor necrosis factor-alpha production by tristetraprolin. *Science.* 281, 1001-1005.

Carballo, E., Lai, W. S. and Blackshear, P. J. (2000). Evidence that tristetraprolin is a physiological regulator of granulocyte-macrophage colony-stimulating factor messenger RNA deadenylation and stability. *Blood.* 95, 1891-1899.

Carrick, D. M., Lai, W. S. and Blackshear, P. J. (2004). The tandem CCCH zinc finger protein tristetraprolin and its relevance to cytokine mRNA turnover and arthritis. *Arthritis Res. Ther.* 6, 248-264.

Ceconi, C., Boraso, A., Mele, D., Guardigli, R. and Ferrari, R. (2004). TNFalpha in patients with congestive heart failure. *Basic. Res. Cardiol.* 99, 12-17.

Ciais, D., Cherradi, N., Bailly, S., Grenier, E., Berra, E., Pouyssegur, J., LaMarre, J. and Feige, J. J. (2004). Destabilization of vascular endothelial growth factor mRNA by the zinc-finger protein TIS11b. *Oncogene.* 23, 8673-8680.

Emoto, M., Anno, T., Sato, Y., Tanabe, K., Okuya, S., Tanizawa, Y., Matsutani, A. and Oka, Y. (2001) Troglitazone treatment increases plasma vascular endothelial growth factor in diabetic patients and its mRNA in 3T3-L1 adipocytes. *Diabetes*. 50, 1166-1170.

Ferrara, N., Gerber, H. P. and LeCouter, J. (2003). The biology of VEGF and its receptors *Nat. Med.* 9, 669-676.

Ferrara, N. and Kerbel, R. S. (2005) Angiogenesis as a therapeutic target. *Nature.* 438, 967-974.

Goldberg-Cohen, I., Furneauxb, H. and Levy, A. P. (2002) .A 40-bp RNA element that mediates stabilization of vascular endothelial growth factor mRNA by HuR. *J. Biol. Chem.* 277, 13635-13640.

Gurantz, D., Cowling, R. T., Villarreal, F. J. and Greenberg, B. H. (1999). Tumor necrosis factor-alpha upregulates angiotensin II type 1 receptors on cardiac fibroblasts. *Circ. Res.* 85, 272-279.

Hikoso, S., Yamaguchi, O., Higuchi, Y., Hirotani, S., Takeda, T., Kashiwase, K., Watanabe, T., Taniike, M., Tsujimoto, I., Asahi, M., Matsumura, Y., Nishida, K., Nakajima, H., Akira, S., Hori, M. and Otsu, K. (2004). Pressure overload induces cardiac dysfunction and dilation in signal transducer and activator of transcription 6-deficient mice. *Circulation.* 110, 2631-2637.

Hotamisligil, G. S. (2006) Inflammation and metabolic disorders. *Nature.* 444, 860-867.

Khanna, D., McMahon, M. and Furst, D. E. (2004). Anti-tumor necrosis factor alpha therapy and heart failure: what have we learned and where do we go from here? *Arthritis Rheum.* 50, 1040-1050.

Lai, W. S. and Blackshear, P. J. (2001). Interactions of CCCH zinc finger proteins with mRNA: tristetraprolin- mediated AU-rich element-dependent mRNA degradation can occur in the absence of a poly(A) tail. *J. Biol. Chem.* 276, 23144-23154.

Lai, W. S., Carballo, E., Strum, J. R., Kennington, E. A., Phillips, R. S. and Blackshear, P. J. (1999). Evidence that tristetraprolin binds to AU-rich elements and promotes the deadenylation and destabilization of tumor necrosis factor alpha mRNA. *Mol. Cell. Biol.* 19, 4311-4323.

Lai, W. S., Carballo, E., Thorn, J. M., Kennington, E. A. and Blackshear, P. J. (2000). Interactions of CCCH zinc finger proteins with mRNA. Binding of tristetraprolin-related zinc finger proteins to Au-rich elements and destabilization of mRNA. *J. Biol. Chem.* 275, 17827-17837.

Lai, W. S., Stumpo, D. J. and Blackshear, P. J. (1990). Rapid insulin-stimulated accumulation of an mRNA encoding a proline-rich protein. *J. Biol. Chem.* 265, 16556-16563.

Levine, B., Kalman, J., Mayer, L., Fillit, H. M. and Packer, M. (1990). Elevated circulating levels of tumor necrosis factor in severe chronic heart failure. *N. Engl. J. Med.* 323, 236-241.

Lim, H. S., Lip, G. Y. and Blann, A. D. (2005) Angiopoietin-1 and angiopoietin-2 in diabetes mellitus: relationship to VEGF, glycaemic control, endothelial damage/dysfunction and atherosclerosis. *Atherosclerosis.* 180, 113-118.

Mann, D. L. (2002). Inflammatory mediators and the failing heart: past, present, and the foreseeable future. *Circ. Res.* 91, 988-998.

Meyer, S., Temme, C. and Wahle, E. (2004). Messenger RNA turnover in eukaryotes: pathways and enzymes. *Crit. Rev. Biochem. Mol. Biol.* 39, 197-216.

Misquitta, C. M., Iyer, V. R., Werstiuk, E. S. and Grover, A. K. (2001). The role of 3'-untranslated region (3'-UTR) mediated mRNA stability in cardiovascular pathophysiology. *Mol. Cell. Biochem.* 224, 53-67.

Miyazawa-Hoshimoto, S., Takahashi, K., Bujo, H., Hashimoto, N., Yagui, K. and Saito, Y. (2005) Roles of degree of fat deposition and its localization on VEGF expression in adipocytes. *Am. J. Physiol Endocrinol. Metab.* 288, E1128-E1136.

Murata, T., Yoshino, Y., Morita, N. and Kaneda, N. (2002). Identification of nuclear import and export signals within the structure of the zinc finger protein TIS11 *Biochem. Biophys. Res. Commun.* 293, 1242-1247.

Nishimura, S., Manabe, I., Nagasaki, M., Hosoya, Y., Yamashita, H., Fujita, H., Ohsugi, M., Tobe, K., Kadowaki, T., Nagai, R. and Sugiura, S. (2007) Adipogenesis in obesity requires close interplay between differentiating adipocytes, stromal cells and blood vessels. *Diabetes.* 56, 1517-1526.

Peng, J., Gurantz, D., Tran, V., Cowling, R. T. and Greenberg, B. H. (2002). Tumor necrosis factor-alpha-induced AT1 receptor upregulation enhances angiotensin II-mediated cardiac fibroblast responses that favor fibrosis. *Circ. Res.* 91, 1119-1126.

Phillips, K., Kedersha, N., Shen, L., Blackshear, P. J. and Anderson, P. (2004). Arthritis suppressor genes TIA-1 and TTP dampen the expression of tumor necrosis factor alpha, cyclooxygenase 2, and inflammatory arthritis. *Proc. Natl. Acad. Sci. U. S. A.* 101, 2011-2016.

Phillips, R. S., Ramos, S. B. and Blackshear, P. J. (2002). Members of the tristetraprolin family of tandem CCCH zinc finger proteins exhibit CRM1-dependent nucleocytoplasmic shuttling. *J. Biol. Chem.* 277, 11606-11613.

Sawaoka, H., Dixon, D. A., Oates, J. A. and Boutaud, O. (2003). Tristetraprolin binds to the 3'-untranslated region of cyclooxygenase-2 mRNA. a polyadenylation variant in a cancer cell line lacks the binding site. *J. Biol. Chem.* 278, 13928-13935.

Schulz, R., Aker, S., Belosjorow, S. and Heusch, G. (2004). TNFalpha in ischemia/reperfusion injury and heart failure. *Basic Res. Cardiol.* 99, 8-11.

Shih, S. C. and Claffey, K. P. (1999). Regulation of human vascular endothelial growth factor mRNA stability in hypoxia by heterogeneous nuclear ribonucleoprotein L. *J. Biol. Chem.* 274, 1359-1365.

Sully, G., Dean, J. L., Wait, R., Rawlinson, L., Santalucia, T., Saklatvala, J. and Clark, A. R. (2004). Structural and functional dissection of a conserved destabilizing element of cyclo-oxygenase-2 mRNA: evidence against the involvement of AUF-1 [AU-rich element/poly(U)-binding/degradation factor-1], AUF-2, tristetraprolin, HuR (Hu antigen R) or FBP1 (far-upstream-sequence-element-binding protein 1). *Biochem. J.* 377, 629-639.

Taylor, G. A., Carballo, E., Lee, D. M., Lai, W. S., Thompson, M. J., Patel, D. D., Schenkman, D. I., Gilkeson, G. S., Broxmeyer, H. E., Haynes, B. F. and Blackshear, P. J. (1996). A pathogenetic role for TNF alpha in the syndrome of cachexia, arthritis, and autoimmunity resulting from tristetraprolin (TTP) deficiency. *Immunity.* 4, 445-454.

von Haehling, S., Jankowska, E. A. and Anker, S. D. (2004). Tumour necrosis factor-alpha and the failing heart--pathophysiology and therapeutic implications. *Basic Res. Cardiol.* 99, 18-28.

Wolfe, F. and Michaud, K. (2004). Heart failure in rheumatoid arthritis: rates, predictors, and the effect of anti-tumor necrosis factor therapy. *Am. J. Med.* 116, 305-11.

Worthington, M. T., Pelo, J. W., Sachedina, M. A., Applegate, J. L., Arseneau, K. O. and Pizarro, T. T. (2002). RNA binding properties of the AU-rich element-binding recombinant Nup475/TIS11/tristetraprolin protein. *J. Biol. Chem.* 277, 48558-48564.

Yu, H., Stasinopoulos, S., Leedman, P. and Medcalf, R. L. (2003). Inherent instability of plasminogen activator inhibitor type 2 mRNA is regulated by tristetraprolin. *J. Biol. Chem.* 278, 13912-13918.

In: Messenger RNA Research Perspectives
Editor: Takeo Takeyama, pp. 17-18
ISBN: 978-1-60021-814-9

Expert Commentary C

Recent Research on the Intracellular Transport and Secretion of Pituitary Hormone

Akira Matsuno
Department of Neurosurgery, Teikyo University Chiba Medical Center,
3426-3 Anesaki, Ichihara City, Chiba 299-0111, Japan

The intracellular transport and secretion of pituitary hormone is an issue of recent interest among researchers. The rab3 protein is involved in the exocytosis of synaptic vesicles and secretory granules in the central nervous system and anterior pituitary cells. We have elucidated the role of rab3B in growth hormone (GH) secretion and the mutual relationship between rab3B and soluble N-ethylmaleimide-sensitive factor attachment protein receptor (SNARE) proteins. Confocal laser scanning microscopic observation of double immunohistochemical stainings for rab3B and GH revealed an increase of immunoreactivity for rab3B in growth hormone-releasing hormone (GHRH)-treated rats and a decrease of immunoreactivity for rab3B in and somatostatin (SRIF)-treated rats. Moreover, confocal laser scanning microscopic observation of double immunohistochemical stainings of synaptosomal-associated protein of 25 kDa (SNAP-25), syntaxin and rab3B revealed co-localization of rab3B and the SNARE proteins in GHRH-treated rats, and dissociation in SRIF-treated rats. These results suggest that rab3B plays an important role in GH secretion from the anterior pituitary cells, and that SNAP-25 and syntaxin act alongside rab3B in the functional regulation of GH secretion. Other reports have indicated that rho and rab proteins such as rab8B and rab27B are involved in the intracellular transport and secretion of pituitary hormones. Functional analyses of rab and rho protein families and the SNARE system are required to elucidate the precise mechanisms of the intracellular transport and secretion of pituitary hormones. Meanwhile, in order to examine the relationship among molecules that play important roles in transportation and secretion of pituitary hormone, the real-time observation in a living cell is essential. Thus, we have developed an experimental pituitary cell line that has secretory granules of GH linked to enhanced yellow fluorescein protein

(EYFP). This stable GH3 cell secretes GH linked to EYFP upon stimulated by Ca2+ influx or Ca2+ release from storage. This GH3 cell is useful for the real-time visualization of the intracellular transport and secretion of GH. The visualization of the process of transcription, translation, transport, and secretion of pituitary hormone is very useful for our insights of pathophysiological studies of pituitary gland.

In: Messenger RNA Research Perspectives
Editor: Takeo Takeyama, pp. 19-26

ISBN: 978-1-60021-814-9

Short Communication A

The Control of mRNA Stability by p38 Mitogen-Activated Protein Kinase and Tristetraprolin

Jonathan L.E. Dean*

Kennedy Institute of Rheumatology Division, Faculty of Medicine, Imperial College London, 1 Apenlea Road, Hammersmith, London, W6 8LH, United Kingdom.

p38 mitogen-activated protein kinase is activated by pro-inflammatory stimuli and is generally recognised as a as a key signalling molecule whose activity is required for the expression of a range of pro-inflammatory mediators. For this reason considerable effort has been invested into the development of inhibitors of p38 and its downstream kinase, MK2, for the therapy of chronic inflammatory diseases such as rheumatoid arthritis.

Our understanding of p38 function has been greatly accelerated by the development of specific small molecule inhibitors of the kinase. These pyridinyl imidazole compounds were found to potently inhibit cytokine production induced by pro-inflammatory stimuli. We and others showed that the mRNAs of inflammatory mediators, such as COX-2 (Dean et al., 1999; Ridley et al., 1998), were destabilised following inhibition of p38. We concluded that p38 MAPK stabilises COX-2 mRNA and together with others went on to show that other pro-inflammatory transcripts are regulated in a similar fashion (Clark et al., 2003). All p38-regulated transcripts contain AU-rich elements in their 3'-untranslated regions (3'-UTR). These elements promote rapid deadenylation and rapid decay of the body of the mRNA. p38 stabilises mRNA by inhibiting ARE-mediated mRNA decay (Lasa et al., 2000; Winzen et al., 1999). p38 specifically inhibits deadenylation rather than blocking the decay of the mRNA body (Dean et al. 2003).

In the case of TNF mRNA, the protein tristetraprolin (TTP) binds to the TNF ARE and targets the mRNA for degradation (Carballo et al., 1998). This protein is induced by pro-inflammatory stimuli and p38 activity is required for its induction (Mahtani et al., 2001). p38

* Jonathan L.E. Dean: Email: jonathan.dean@imperial.ac.uk

activates the downstream kinase, MK2, and this phosphorylates TTP (Chrestensen et al., 2004; Mahtani et al., 2001) and inactivates its destabilising function (Stoecklin et al., 2004). It is then held in an inactive state, poised to trigger rapid decay of TNF mRNA as soon as p38/MK2 activity subsides. Thus, regulation of the inflammatory response by p38 is not as simple as originally believed. In addition to activating the induction of inflammatory genes, p38 activity is also needed to prime cells for the resolution of the expression of these genes.

Recent Progress in the Mechanism of Induction of Inflammatory Response Genes via Stabilisation of mRNA by p38 MAPK: The Involvement of TTP

We found that TTP is phosphorylated by MK2 (Mahtani et al., 2001), suggesting that TTP was possibly the missing link between p38 and inflammatory mRNAs. The MK2 phosphorylation sites have been identified as serines 52 and 178 (Chrestensen et al., 2004). Blackshear and co-workers have also reported that LPS-induced TNF protein production by $TTP^{-/-}$ macrophages is unaffected by p38 inhibition (Carballo et al., 2001). However, another group found that the expression of TNF by $TTP^{-/-}$ macrophages is sensitive to p38 blockade (Kontoyiannis et al., 2001). More recently, SB203580 was found to have only a weak effect on TNF protein expression in $TTP^{-/-}$ macrophages (Hitti et al., 2006). The creation of $MK2^{-/-}$ $TTP^{-/-}$ mice by Hitti et al. has also helped resolve this issue. Macrophages from these mice no longer display decreased TNF production (Hitti et al., 2006) as in the case of MK2-null macrophages (Kotlyarov et al., 1999). Rather they overproduce TNF, as seen in the TTP knockout cells (Hitti et al., 2006). This provides additional support for the idea that in macrophages, at least, TTP is responsible for p38-mediated mRNA stabilisation.

TNF mRNA has reduced stability in $MK2^{-/-}$ macrophages, but it is stable in $MK2^{-/-}$ $TTP^{-/-}$ cells (Hitti et al., 2006). We have recently found that destabilisation of TNF mRNA upon p38 blockade is largely ablated in $TTP^{-/-}$ cells (JLED, manuscript submitted). Taken together these findings indicate that TTP is required for the increased decay of TNF mRNA that occurs in the absence of p38/MK2 activity.

It now appears clear that TTP lies downstream of MK2. It is thus likely that MK2 can inactivate TTP's mRNA destabilising function. To date the strongest experimental evidence of this comes from a study in which a TNF reporter mRNA and TTP were coexpressed in COS7 cells (Stoecklin et al., 2004). The reporter mRNA bearing the TNF 3'UTR was found to be relatively stable in the absence of TTP overexpression. Expression of exogenous wild-type TTP was found to destabilise the reporter mRNA and this was largely reversed by co-expression of an active mutant of MK2 (Stoecklin et al., 2004). This mutant kinase was unable to inactivate mRNA destabilisation directed by mutant forms of TTP in which the MK2 phosphorylation sites were rendered non-phosphorylatable by mutation to alanine (Stoecklin et al., 2004). However, others have performed similar experiments and have failed to show inactivation of TTP upon activation of the p38 pathway by an active mutant of MKK6 (Rigby et al., 2005).

TTP is not the only TTP family member whose destabilising activity is inhibited by phosphorylation. The activity of the TTP homologue, BRF1, is blocked by protein kinase B-mediated phosphorylation (Benjamin et al., 2006; Schmidlin et al., 2004). In this case direct evidence for the inactivation of the ARE-BP by phosphorylation has been obtained. Degradation of RNA substrates in vitro directed by the addition of recombinant BRF1 is inhibited when the ARE-BP is phosphorylated by PKB prior to its addition (Schmidlin et al., 2004). It would be of great interest to reconstitute regulation of mRNA decay involving TTP and p38 in vitro. This would permit a direct test of whether MK2 directly inactivates TTP.

At present it is unclear precisely how MK2 inactivates the destabilising function of TTP. One proposed mechanism is that phosphorylation of TTP modulates its affinity for AREs. Dephosphorylated TTP has been suggested to bind RNA probes with higher affinity than the phosphorylated protein (Carballo et al., 2001). This has also been suggested to occur specifically in response to phosphorylation of TTP by MK2 (Hitti et al., 2006). However, others have found that phosphorylation of serines 52 and 178 had no effect on the affinity of TTP for RNA (Sun et al., 2006).

TTP is thought to target mRNAs for rapid degradation by recruiting mRNA decay enzymes to ARE-containing mRNAs. In particular, TTP has been shown to associate with the multi-component exonuclease complex, the exosome, the 5'-3' exonuclease, Xrn1, the deadenylase CCR4, and the decapping enzyme hDcp2 (Lykke-Andersen and Wagner, 2005). Another proposed mechanism is that phosphorylation of TTP prevents it from recruiting mRNA decay factors. Phosphorylation of TTP on serines 52 and 178 creates binding sites for 14-3-3 (Chrestensen et al., 2004; Johnson et al., 2002; Stoecklin et al., 2004). Activation of the p38 pathway promotes TTP:14-3-3 complex formation and this has been proposed to prevent TTP from interacting with mRNA decay factors (Stoecklin et al., 2004). Inactivation of the destabilising function of the ARE-BP KSRP, via phosphorylation by PKB, involves 14-3-3 binding (Gherzi et al., 2006). In this case, 14-3-3 was found to inhibit the recruitment of the exosome by KSRP (Gherzi et al., 2006). It will be interesting to find out if TTP is regulated in a similar fashion.

At present it is not known whether TTP is responsible for p38-mediated regulation of mRNA stability in cells other than macrophages and in response to other pro-inflammatory stimuli. TTP is expressed in other cell types. It is induced by serum in murine embryonic fibroblasts and regulates a number of transcripts in these cells, including Ier3 (Lai et al., 2006). It is also induced in T cells upon T cell receptor and CD28 coreceptor ligation and regulates IL-2 mRNA stability (Ogilvie et al., 2005). It is not known whether TTP is involved in p38-mediated regulation induced by IL-1 in cells such as human fibroblasts and HeLa cells. TTP can be detected in IL-1-treated HeLa cells, although only a very small amount is expressed in these cells (A. Clark and C. Asensio, personal communication). It is unclear whether it is induced by IL-1 in human fibroblasts. It is possible that homologues of TTP, such as BRF1 and BRF2, which are abundant in these cells may fulfil the role of TTP.

Another important question is whether mRNAs other than that for TNF are co-regulated by p38 and TTP. We have recently discovered that four other mRNAs (COX-2, IL-6, IL-10 and IL-1α) are regulated by both p38 and TTP (JLED, in preparation). We found that the rapid decay of these mRNAs initiated by addition of p38 inhibitor no longer occurred in TTP-deficient macrophages. Thus TTP may act downstream of p38 to regulate many, if not all, p38-regulated inflammatory mediator mRNAs.

Resolution of the Inflammatory Response by P38 and TTP

TTP was originally found as a gene whose expression is induced by TPA (or phorbol ester), hence its other name (TIS11: TPA-inducible sequence 11). TTP has been long known to be induced by LPS. We showed that the induction of the protein by LPS requires p38 activity (Mahtani et al., 2001). Andy Clark's group have gone on to show that p38 exerts tight control over the expression of TTP protein through regulation of TTP mRNA stability (Tchen et al., 2004) and TTP protein turnover (Brook et al., 2006).

Inhibition of p38 causes a decrease in TTP mRNA in LPS-treated macrophages (Mahtani et al., 2001). This is in part due to stabilisation of TTP mRNA by MK2 (Tchen et al., 2004). The induction of TTP mRNA in response to LPS is reduced in MK2$^{-/-}$ macrophages (Hitti et al., 2006). TTP mRNA is also less stable in these cells (Hitti et al., 2006). TTP protein can bind to the ARE in the 3'-UTR of its own mRNA in vitro and this may allow for auto-regulation (Tchen et al., 2004). TTP mRNA has also been co-immunoprecipitated with TTP protein (Brooks et al., 2004). However, in embryonic fibroblasts from TTP-null mice the chimaeric TTP mRNA containing the neomycin insert displays identical stability to TTP mRNA in wild type cells (Lai et al., 2006). In this study, serum was used as the stimulus to induce TTP (Lai et al., 2006), it is possible that TTP autoregulation is stimulus and cell-type specific.

p38 also regulates TTP protein stability. In experiments where cycloheximide was used to block translation, TTP protein was found to decay rapidly in the presence of p38 inhibitor SB202190 (Brook et al., 2006). TTP protein also decays more rapidly in MK2$^{-/-}$ macrophages than in wild-type cells (Hitti et al., 2006) and TTP protein stability is dependent on the integrity of serines 52 and 178 (Brook et al., 2006). The stability of the protein is further reduced by the addition of U0126, an ERK pathway inhibitor together with SB202190 (Brook et al., 2006). The degradation of TTP protein that occurs when p38 activity drops has been suggested to prevent the induction of apoptosis that happens as a result of prolonged expression of TTP (Brook et al., 2006). It may also be vital to ensure that TTP only targets AU-rich mRNAs for decay for a brief, finite period after which further stimulation of the cells can again induce inflammatory gene expression. Nucleocytoplasmic shuttling of TTP has also been suggested to serve similar functions. Phosphorylation of TTP by MK2 results in cytoplasmic localisation of the ARE-BP (Brook et al., 2006). It appears that MK2 phosphorylation causes retention of TTP in the cytoplasm involving 14-3-3 binding (Johnson et al., 2002).

p38 phosphorylates and activates MK2 and this in turn induces TTP mRNA and phosphorylates TTP protein. TTP is thereby rendered inactive and inflammatory mRNAs (e.g. TNF mRNA) are stable. TTP is held in a latent, stable and inactivated state in the cytoplasm. When activity in the p38 pathway subsides TTP is dephosphorylated and inflammatory mRNAs, such as TNF mRNA are targeted for degradation. Dephosphorylated TTP protein itself is also degraded and TTP is imported into the nucleus.

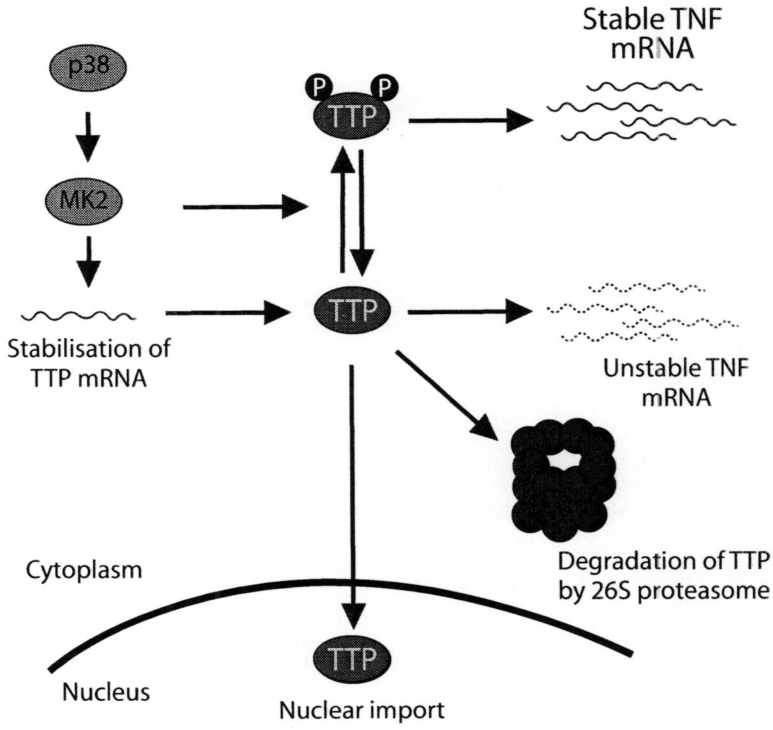

Figure 1. Regulation of TTP by the p38 pathway.

Conclusion

In the light of these findings it is perhaps inappropriate to regard the function of p38, at the post-transcriptional level, as simple induction of the inflammatory response. It must be recognised as important for both the induction and the resolution of inflammatory gene expression. This brings into question the usefulness of targeting p38 for the treatment of chronic inflammatory diseases. The combined effect of p38 inhibitors on inflammatory gene expression in increasing the activity of TTP, but also in reducing its expression could be neutral. In fact, in primary murine macrophages unlike acute inhibition of p38, where the inhibitor is added to the cells after a period of stimulation resulting in TNF mRNA destabilisation, continuous blockade of p38 activity has no effect on the TNF mRNA half-life (JLED, in preparation). This may explain why, surprisingly, LPS-induced steady-state TNF mRNA levels change little in murine macrophages in response to p38 inhibition. Drugs that specifically activate TTP rather than blocking its expression may be more effective in disease therapy than inhibitors of p38/MK2.

ACKNOWLEDGEMENTS

I am grateful to Andy Clark for reading the manuscript and to my PhD student Corina Tudor for performing the unpublished work.

REFERENCES

Benjamin, D., Schmidlin, M., Min, L., Gross, B. and Moroni, C. (2006) BRF1 protein turnover and mRNA decay activity are regulated by protein kinase B at the same phosphorylation sites. *Mol. Cell Biol*, 26, 9497-9507.

Brook, M., Tchen, C.R., Santalucia, T., McIlrath, J., Arthur, J.S., Saklatvala, J. and Clark, A.R. (2006) Posttranslational regulation of tristetraprolin subcellular localization and protein stability by p38 mitogen-activated protein kinase and extracellular signal-regulated kinase pathways. *Mol. Cell Biol*, 26, 2408-2418.

Brooks, S.A., Connolly, J.E. and Rigby, W.F. (2004) The role of mRNA turnover in the regulation of tristetraprolin expression: evidence for an extracellular signal-regulated kinase-specific, AU-rich element-dependent, autoregulatory pathway. *J. Immunol*, 172, 7263-7271.

Carballo, E., Cao, H., Lai, W.S., Kennington, E.A., Campbell, D. and Blackshear, P.J. (2001) Decreased sensitivity of tristetraprolin-deficient cells to p38 inhibitors suggests the involvement of tristetraprolin in the p38 signaling pathway. *J. Biol. Chem*, 276, 42580-42587.

Carballo, E., Lai, W.S. and Blackshear, P.J. (1998) Feedback inhibition of macrophage tumor necrosis factor-alpha production by tristetraprolin. *Science*, 281, 1001-1005.

Chrestensen, C.A., Schroeder, M.J., Shabanowitz, J., Hunt, D.F., Pelo, J.W., Worthington, M.T. and Sturgill, T.W. (2004) MAPKAP kinase 2 phosphorylates tristetraprolin on in vivo sites including Ser178, a site required for 14-3-3 binding. *J. Biol. Chem*, 279, 10176-10184.

Clark, A.R., Dean, J.L. and Saklatvala, J. (2003) Post-transcriptional regulation of gene expression by mitogen-activated protein kinase p38. *FEBS Lett*, 546, 37-44.

Dean, J.L., Brook, M., Clark, A.R. and Saklatvala, J. (1999) p38 mitogen-activated protein kinase regulates cyclooxygenase-2 mRNA stability and transcription in lipopolysaccharide-treated human monocytes. *J. Biol. Chem*, 274, 264-269.

Gherzi, R., Trabucchi, M., Ponassi, M., Ruggiero, T., Corte, G., Moroni, C., Chen, C.Y., Khabar, K.S., Andersen, J.S. and Briata, P. (2006) The RNA-Binding Protein KSRP Promotes Decay of beta-Catenin mRNA and Is Inactivated by PI3K-AKT Signaling. *PLoS Biol*, 5, e5.

Hitti, E., Iakovleva, T., Brook, M., Deppenmeier, S., Gruber, A.D., Radzioch, D., Clark, A.R., Blackshear, P.J., Kotlyarov, A. and Gaestel, M. (2006) Mitogen-activated protein kinase-activated protein kinase 2 regulates tumor necrosis factor mRNA stability and translation mainly by altering tristetraprolin expression, stability, and binding to adenine/uridine-rich element. *Mol. Cell Biol*, 26, 2399-2407.

Johnson, B.A., Stehn, J.R., Yaffe, M.B. and Blackwell, T.K. (2002) Cytoplasmic localization of tristetraprolin involves 14-3-3-dependent and -independent mechanisms. *J. Biol. Chem*, 277, 18029-18036.

Kontoyiannis, D., Kotlyarov, A., Carballo, E., Alexopoulou, L., Blackshear, P.J., Gaestel, M., Davis, R., Flavell, R. and Kollias, G. (2001) Interleukin-10 targets p38 MAPK to modulate ARE-dependent TNF mRNA translation and limit intestinal pathology. *Embo J*, 20, 3760-3770.

Kotlyarov, A., Neininger, A., Schubert, C., Eckert, R., Birchmeier, C., Volk, H.D. and Gaestel, M. (1999) MAPKAP kinase 2 is essential for LPS-induced TNF-alpha biosynthesis. *Nat. Cell Biol*, 1, 94-97.

Lai, W.S., Parker, J.S., Grissom, S.F., Stumpo, D.J. and Blackshear, P.J. (2006) Novel mRNA targets for tristetraprolin (TTP) identified by global analysis of stabilized transcripts in TTP-deficient fibroblasts. *Mol. Cell Biol*, 26, 9196-9208.

Lasa, M., Mahtani, K.R., Finch, A., Brewer, G., Saklatvala, J. and Clark, A.R. (2000) Regulation of cyclooxygenase 2 mRNA stability by the mitogen-activated protein kinase p38 signaling cascade. *Mol. Cell Biol*, 20, 4265-4274.

Lykke-Andersen, J. and Wagner, E. (2005) Recruitment and activation of mRNA decay enzymes by two ARE-mediated decay activation domains in the proteins TTP and BRF-1. *Genes Dev*, 19, 351-361.

Mahtani, K.R., Brook, M., Dean, J.L., Sully, G., Saklatvala, J. and Clark, A.R. (2001) Mitogen-activated protein kinase p38 controls the expression and posttranslational modification of tristetraprolin, a regulator of tumor necrosis factor alpha mRNA stability. *Mol. Cell Biol*, 21, 6461-6469.

Ogilvie, R.L., Abelson, M., Hau, H.H., Vlasova, I., Blackshear, P.J. and Bohjanen, P.R. (2005) Tristetraprolin down-regulates IL-2 gene expression through AU-rich element-mediated mRNA decay. *J. Immunol*, 174, 953-961.

Ridley, S.H., Dean, J.L., Sarsfield, S.J., Brook, M., Clark, A.R. and Saklatvala, J. (1998) A p38 MAP kinase inhibitor regulates stability of interleukin-1-induced cyclooxygenase-2 mRNA. *FEBS Lett*, 439, 75-80.

Rigby, W.F., Roy, K., Collins, J., Rigby, S., Connolly, J.E., Bloch, D.B. and Brooks, S.A. (2005) Structure/function analysis of tristetraprolin (TTP): p38 stress-activated protein kinase and lipopolysaccharide stimulation do not alter TTP function. *J. Immunol*, 174, 7883-7893.

Schmidlin, M., Lu, M., Leuenberger, S.A., Stoecklin, G., Mallaun, M., Gross, B., Gherzi, R., Hess, D., Hemmings, B.A. and Moroni, C. (2004) The ARE-dependent mRNA-destabilizing activity of BRF1 is regulated by protein kinase B. *Embo J*, 23, 4760-4769.

Stoecklin, G., Stubbs, T., Kedersha, N., Wax, S., Rigby, W.F., Blackwell, T.K. and Anderson, P. (2004) MK2-induced tristetraprolin:14-3-3 complexes prevent stress granule association and ARE-mRNA decay. *Embo J*, 23, 1313-1324.

Sun, L., Stoecklin, G., Van Way, S., Hinkovska-Galcheva, V., Guo, R.F., Anderson, P. and Shanley, T.P. (2006) TTP/14-3-3 complex formation protects TTP from dephosphorylation by protein phosphatase 2A and stabilizes TNF-alpha mRNA. *J. Biol. Chem*.

Tchen, C.R., Brook, M., Saklatvala, J. and Clark, A.R. (2004) The stability of tristetraprolin mRNA is regulated by mitogen-activated protein kinase p38 and by tristetraprolin itself. *J. Biol. Chem*, 279, 32393-32400.

Winzen, R., Kracht, M., Ritter, B., Wilhelm, A., Chen, C.Y., Shyu, A.B., Muller, M., Gaestel, M., Resch, K. and Holtmann, H. (1999) The p38 MAP kinase pathway signals for cytokine-induced mRNA stabilization via MAP kinase-activated protein kinase 2 and an AU-rich region-targeted mechanism. *Embo J*, 18, 4969-4980.

In: Messenger RNA Research Perspectives
Editor: Takeo Takeyama, pp. 27-33

ISBN: 978-1-60021-814-9

Short Communication B

Gene Expression of Enamel-Related Proteins in Rat Odontoblasts

T. Nagano**[*1]**, S. Oida**[2]**, H. Yamaguchi**[1]**, H. Sumimoto**[1]**,
***T. Karakida**[2]**, K. Gomi**[1]**, T. Arai**[1]** and M. Fukae**[1]*

[1] Department of Periodontics and Endodontics, School of Dental Medicine, Tsurumi University, 2-1-3 Tsurumi, Tsurumi-ku, Yokohama 230-8501, Japan

[2] Department of Biochemistry, School of Dental Medicine, Tsurumi University, 2-1-3 Tsurumi, Tsurumi-ku, Yokohama 230-8501, Japan

ABSTRACT

Objectives: Amelogenin, enamelin, sheathlin (ameloblastin/amelin), enamelysin, and KLK4 are the major structural proteins and proteinases in developing tooth enamel. Recently, it has been shown odontoblasts express the amelogenin. In this study, we examined whether or not the other enamel proteins and proteinases were expressed in the rat odontoblasts.

Methods: We prepared secretory-stage enamel organ epithelia (EOE), maturation-stage EOE, and odontoblasts lining dentin under the secretory and maturation stages enamel from the labial site of the developing incisors. Total RNA of each cell layer sample was obtained with the Total RNA Miniprep kit (Stratagene, USA). Single-strand cDNA was prepared from 3μg of each total RNA sample using a Ready to go T-primed First-strand beads kit (Pharmacia, USA). Each of the cDNA was used for PCR. In addition, rat incisor was fixed by Karnovsky's solution. After dehydration, the specimens were embedded in Technovit 7200VLC (Kulzer, Germany) and cut by Exact cutting machine (Exact, Germany). The sections were polished until 50μm and double stained with Stevenel's blue and Van Gieson picro-fuchsin.

Results: Histological examination was revealed that the highly mineralized enamel existed as a narrow layer along the enamel-dentin junction. The mRNAs encoding the

* Corresponding author: Takatoshi Nagano; Department of Periodontics and Endodontics, School of Dental Medicine, Tsurumi University, 2-1-3 Tsurumi, Tsurumi-ku, Yokohama 230-8501, Japan; Fax: +81-45-573-9599; E-mail : nagano-takatoshi@tsurumi-u.ac.jp

structural enamel proteins and proteinases were found more or less in odontoblast samples. The amelogenin, enamelin, and sheathlin mRNAs were expressed odontoblasts under the secretory stage enamel. The enamelysin and KLK4 mRNAs were expressed both EOE and odontoblasts. The enamelysin mRNA was found in the secretory EOE. For KLK4, this trend was reversed. The KLK4 mRNA expression in maturation EOE was higher in secretory EOE. Furthermore, the KLK4 mRNA was expressed more in odontoblasts than in the secretory EOE.

Conclusion: These results indicate that the rat odontoblasts express enamel proteins and proteinases mRNA, which may play some part in the formation of dentin.

Keywords: amelogenin, enamelysin, KLK4, odontoblasts, ameloblasts.

INTRODUCTION

Amelogenin, enamelin, sheathlin (ameloblastin/amelin), enamelysin (MMP-20), and KLK4 (EMSP-1) are the major structural proteins and proteinases in developing tooth enamel. Amelogenin is the most abundant enamel protein during the secretory-stage of amelogenesis (Termine et al., 1980). Enamelin is the largest glycoprotein secreted into the matrix (Hu et al., 1997b), while sheathlin cleavage products (Fukae and Tanabe, 1987) concentrate in the sheath space between the rod and interrod enamel (Uchida et al., 1995, Hu et al., 1997a). Amelin (Cerny et al., 1996) and ameloblastin (Krebsbach et al., 1996), first cloned from rat-tooth specific cDNA libraries, are homologs of porcine sheathlin. There are two proteinases, enamelysin (MMP-20) (Bartlett et al., 1996, Fukae et al., 1998) and Kallikrein-4 (KLK4) also known as enamel matrix serine proteinase-1 (EMSP-1) (Simmer et al., 1998), involved in the degradation of amelogenin. Enamelysin participates in the conversion from 25kDa amelogenin to 20kDa amelogenin (Fukae and Tanabe, 1998). KLK4 is a proteolytic enzyme that was first detected in porcine enamel (Fukae et al., 1977), and degrades the 20kDa amelogenin into two fragments, corresponding to the 6kDa and 13kDa amelogenins (Shimizu and Fukae, 1983).

Recently, it has been shown porcine odontoblasts express the amelogenin or enamel-related proteins (Oida et al., 2002 / Nagano et al., 2003).

In this study, we examined whether or not the other enamel proteins and proteinases were expressed in the rat odontoblasts.

METHODS

All experimental procedures involving the use of animals was reviewed and approved by the Institutional Animal Care Program at the Tsurumi University.

We prepared secretory-stage enamel organ epithelia (EOE) (SA), maturation-stage EOE (MA), and odontoblasts lining dentin under the secretory (YO) and maturation stages enamel (MO) from the labial site of the developing incisors (figure 1).

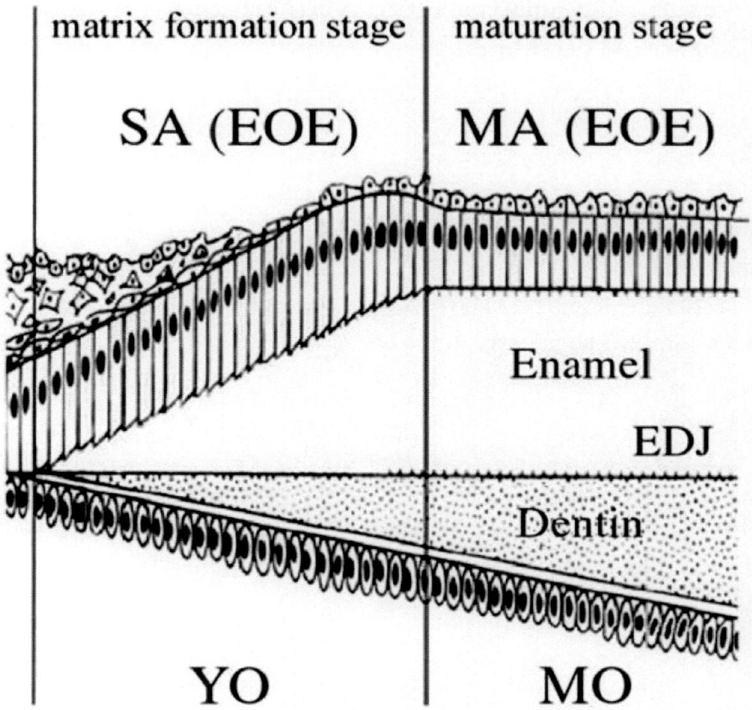

Figure 1. Schematic drawing of ameloblast and odontoblast tissue preparations. SA: secretory EOE, MA: maturation EOE, YO: young odontoblast layer, MO: mature odontoblast layer.

Total RNA of each cell layer sample was obtained with the Total RNA Miniprep kit (Stratagene, U.S.A). Single-strand cDNA was prepared from 3μg of each total RNA sample using a Ready to go T-primed First-strand beads kit (Pharmacia, U.S.A). Specific primers of enamel-related proteins were designed based on porcine and mouse mRNA sequences. Each of the cDNA was used for PCR. PCR conditions (Perkin-Elmer/GeneAmp PCR system 9600) were as follows: PCR started with a 10 min denaturation at 94 ℃, followed by 25 cycles with denaturation at 94 □ for 30 sec, primer annealing at 55 □ for 30 sec, and product extension at 72 □ for 30 sec; in the final cycle, the 72 □ extension was for 7 min. PCR products were analyzed by 4.5% polyacrylamide gel electrophoresis in TBE buffer (pH 8.0). Gels were stained with ethidium bromide, and the bands were visualized under UV light. Specific primers of enamel proteins and proteinases were designed based on porcine and mouse mRNA sequences. The amelogenin sense and anti-sense primers were: 5'ACCCCTCTGAAGTGGTACCAG 3' and 5' TGTTGGGTTGGAGTCATGGAG 3', which generated a 236 bp amplification product. The enamelin sense and anti-sense primers were: 5' AAGAACTGCTGGCCTTACTCC 3' and 5' TCCCCAAAGAATGTTTGGTTG 3', which generated a 433 bp amplification product. The sheathlin sense and anti-sense primers were: 5' CAGAGGTAGCATCTGGGTGTCC 3' and 5' CTCTCAGGGCTCTTGGAAACG 3', which generated a 320 bp amplification product. The enamelysin sense and anti-sense primers were: 5' ATACGTGCAGCGAATAGATGC 3' and 5' CTATTTAGCAACCAATCCAGG 3', which generated a 290 bp amplification product. The

KLK4 sense and anti-sense primers were: 5' ACCATGGGCGAGGACTGCAA 3' and 5' GTTAACTGGCCTGGATGGTCG 3', which generated a 673 bp amplification product. A primer set amplifying glyceraldehyde-3-phosphate dehydrogenase (GAPDH) (Clontech, U.S.A) was used as a control.

Additionally, the PCR products were cloned into pBluescriptIISK(+) (Stratagene, U.S.A), and their nucleotide sequences were determined by cycle sequencing.

In addition, rat incisor was fixed by Karnovsky's solution. After dehydration, the specimens were embedded in Technovit 7200VLC (Kulzer, Germany) and cut by Exact cutting machine (Exact, Germany). The sections were polished until 50μm and double stained with Stevenel's blue and Van Gieson picro-fuchsin

RESULTS

Histological examination was revealed that the highly mineralized enamel (HME) existed as a narrow layer along the enamel-dentin junction (EDJ) (figure 2).

Figure 2. The rat incisor sample contained cells from the inner enamel epithelium, stratum intermedium, odontoblasts, and dental papilla cells. The HME existed as a narrow layer along the EDJ.

The mRNAs encoding the structural enamel proteins and proteinases were found more or less in odontoblast samples (figure 3). The amelogenin, enamelin, and sheathlin mRNAs were expressed odontoblasts under the secretory stage enamel. The enamelysin and KLK4 mRNAs were expressed both EOE and odontoblasts. The enamelysin mRNA was found in the secretory EOE. For KLK4, this trend was reversed. The KLK4 mRNA expression in maturation EOE was higher in secretory EOE. Furthermore, the KLK4 mRNA was expressed more in odontoblasts than in the secretory EOE.

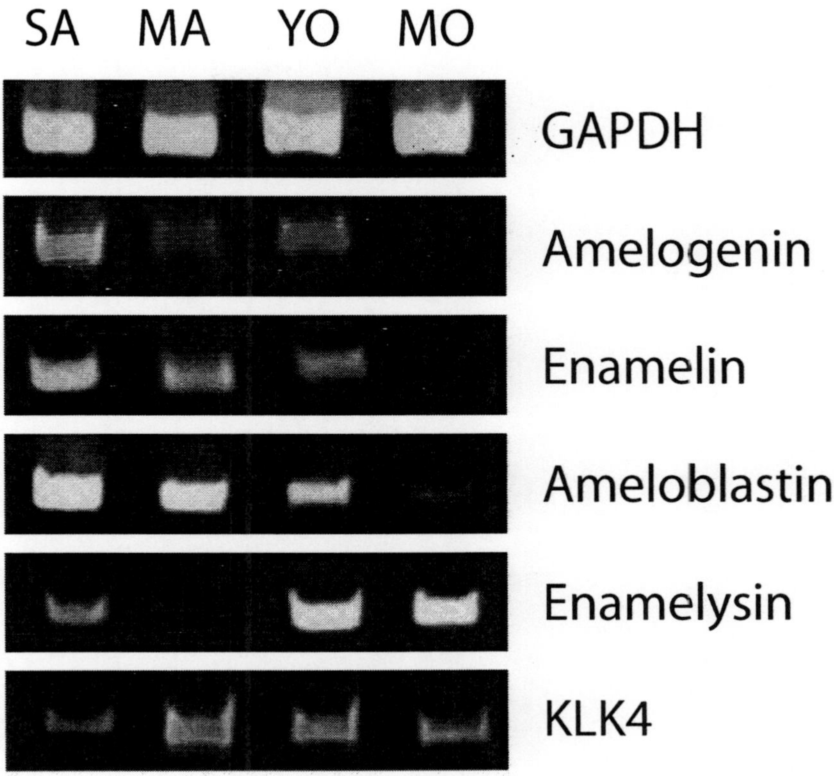

Figure 3. Detection of mRNAs encoding GAPDH, amelogenin, enamelin, sheathlin, enamelysin, and KLK4 after 25 cycles of RT-PCR. PCR products were amplified from the cDNA template and separated by electophoresis on a 4.5% polyacrylamide gel. SA: secretory EOE, MA: maturation EOE, YO: young odontoblast layer, MO: mature odontoblast layer.

Discussion

All structural enamel proteins were expressed in both the EOE and odontoblast cell layer samples. Enamelysin and KLK4 were expressed by both ameloblasts and odontoblasts. In EOE samples, the enamelysin mRNA was found only in the secretory EOE. Enamelysin mRNA levels in secretory EOE were about half that of the odontoblasts, suggesting that secretory ameloblasts specifically may express enamelysin mRNA. For KLK4, this trend was reversed. The KLK4 mRNA expression in maturation EOE was higher in secretory EOE, suggesting that maturation ameloblasts specifically may express KLK4 mRNA higher than odontoblasts. In contrast, KLK4 mRNA in secretory-stage EOE was much lower than in younger and older odontoblasts, indicating that KLK4 mRNA is expressed more in odontoblasts than in the secretory ameloblasts, These results indicate that the rat odontoblasts express enamel proteins and proteinases mRNA, which may play some part in the formation of dentin, particulary HME forming. Enamelysin and KLK4 are expressed by both ameloblasts and odontoblasts. Recently, KLK4 activity was detected in the HME at the EDJ during the secretory stage (Fukae et al., 2002), despite the fact that no KLK4 activity was observed in the inner-layer enamel (Fukae et al., 2001), pre-dentin or dentin (Fukae et al.,

2002). KLK4 activity in the deepest inner layer enamel during the secretory stage is probably secreted through the odontoblast dental tubules at the EDJ where it degrades organic matrix to be replaced by mineral.

Acknowledgements

We thank all the members of our labs, especially Dr. Mikimoto Kanazashi, Dr. Satoshi Shirakawa, Dr. Takako Ishio, Dr. Takako Tanabe, and Dr. Yasuo Yamakoshi. This study was supported by grant-in-aid (No.12671818) for funding High Technology Research Centers from Ministry of Education, Culture, Sports, Science, and Technology of Japan. This work partially supported by funding from the Bio-Venture Project from the Japanese Ministry of Education, Culture, Sports, Science, and Technology.

References

Bartlett JD, Simmer JP, Xue J, Margolis HC, Moreno EC (1996). Molecular cloning and mRNA tissue distribution of a novel matrix metallo proteinase isolated from porcine enamel organ. *Gene* 183:123-128.

Cerny R, Slaby I, Hammarstrom L, Wultz T (1996). A novel gene expressed in rat ameloblasts codes for proteins with cell binding domains. *J. Bone Miner Res* 11:883-891.

Fukae M, Tanabe T, Shimizu M (1977). Proteolytic enzyme activity in porcine immature enamel. *Tsurumi Univ. Dent. J.* 3:15-17.

Fukae M, Tanabe T (1987). Nonamelogenin components of porcine enamel in the protein fraction free from the enamel crystals. *Calcif. Tissue Int.* 40:286-293.

Fukae M, Tanabe T, Uchida T, Lee SK, Ryu OH, Murakami C, *et al.* (1998). Enamelysin (Matrix Metalloproteinase-20) : Localization in the developing tooth and effects of pH and calcium on amelogenin hydrolysis. *J. Dent. Res.* 77:1580-1588.

Fukae M, Tanabe T (1998). Degradation of enamel matrix proteins in porcine secretory enamel. *Connective Tissue Res.* 39:123-129.

Fukae M, Tanabe T, Yamakoshi Y, Yamada M, Ujiie Y, Oida S (2001). Immunoblot detection and expression of enamel proteins at the apical portion of the forming root in porcine permanent incisor tooth germs. *J. Bone Miner Metab.* 19:236-243.

Fukae M, Tanabe T, Nagano T, Ando H, Yamakoshi Y, Yamada M *et al.* (2002). Odontoblast enhance the maturation of enamel crystals by secreting EMSP-1 at the enamel-dentin junction. *J. Dent. Res.* 81:668-672.

Hu C-C, Fukae M, Uchida T, Qian Q, Zhang CH, Ryu OH, *et al.* (1997a). Cloning, cDNA/polypeptide seqences, and immunolocalization of porcine enamel sheath proteins. *J. Dent. Res.* 76:648-657.

Hu C-C, Fukae M, Uchida T, Qian Q, Zhang CH, Ryu OH, *et al.* (1997b). Cloning and characterization of porcine enamelin mRNAs. *J. Dent. Res.* 76:1720-1729.

Krebsbach PH, Lee SK, Matsuki Y, Kozack CA, Yamada K, Yamada Y (1996). Full-length sequence, localization, and chromosomal mapping of ameloblastin. A novel tooth specific gene. *J.Biol. Chem.* 271:4431-4435.

Nagano T, Oida S, Ando H, Gomi K, Arai T, Fukae M (2003). Relative levels of mRNA encoding enamel proteins in enamel organ epithelia and odontoblasts. *J. Dent. Res*, 82:982-986.

Oida S, Nagano T, Yamakoshi Y, Ando H, Yamada M, Fukae M (2002). Amelogenin gene expression in porcine odontoblasts. *J. Dent. Res*, 81:103-108.

Shimizu M, Fukae M (1983). Enamel proteins. In mechanisms of tooth enamel formation, (ed. by Suga S). *Quintessence Publishing Co Inc.,Tokyo*:125-141

Simmer JP, Fukae M, Tanabe T, Yamakoshi Y, Uchida T, Xue J, *et al.* (1998). Purification, characterization, and cloning of enamel matrix serine proteinase 1. *J. Dent. Res.* 77:377-386.

Termine JD, Belcourt AB, Christner PJ, Conn KM, Nylen MU (1980). Properties of dissociatively extracted fetal tooth matrix proteins. *J. Biol. Chem.* 20:9760-9768.

Uchida T, Fukae M, Tanabe T, Yamakoshi Y, Satoda T, Murakami C, *et al.* (1995). Immunochemical and Immunocytochemical study of a 15kDa non-amelogenin and related proteins in the porcine immature enamel: proposal of a new group of enamel proteins 'sheath proteins'. *Biomedical Res.* 16:131-140.

In: Messenger RNA Research Perspectives
Editor: Takeo Takeyama, pp. 35-79
ISBN: 978-1-60021-814-9

Chapter I

Imprint of Early tRNA Diversification on the Genetic Code: Domains of Contiguous Codons Read by Related Adaptors for Sibling Amino Acids[1]

Brian K. Davis*

Research Foundation of Southern California, Inc.
P.O. Box 13595; La Jolla, CA 92039

Abstract

Diverse structural regularities identified in the genetic code, over the last four decades, have been unified with a model of code evolution that equated the time-order of amino acid (aa) addition to the code with the number of reaction steps (from the citrate cycle) required for its synthesis. To help establish the mechanism that coordinated growth of aa synthesis pathways with code formation, evidence was sought that tRNA adaptor molecules functioned as cofactors in aa synthesis, prior to appearance of the last common ancestor (LCA).

Phylogenetic analysis of the conserved trace of ancestral tRNA revealed: (1) LCA tRNA species that descended from a common ancestor contained similar anticodons and were adaptors for aa synthesized from a common precursor. (2) LCA tRNA adaptors for sibling aa generally possessed the same core structure group and had elevated sequence identity. (3) Five domains of contiguous codons read by tRNA, sharing the same core group and bearing sibling aa, were spread along code rows. (4) These coding domains covered most of the code. They also overlapped extensively with code regions assigned to same-family aa and with regions read by phylogenetically-related LCA tRNA. (5) Type ID tRNA were adaptors for five aa, including three NH_4^+ fixers, synthesized on 1-2 step paths. A tRNA-ID was found to be ancestral to the type IA tRNA asparagine-adaptor, which, in turn, displayed kinship with adaptors for fourteen longer-path (4-14

---

[1] Some results presented in this chapter appear in Ref. 52.

* Brian K. Davis: email: davis@resfdnsca.org

steps) aa. (6) Nearly all extant intermediates in aa synthesis, in accord with the foregoing results, were noted to contain a free tRNA attachment site.

Based on these observations, tRNA-dependent aa synthesis was credited with producing code domains by channeling nearest-neighbor codons, in code rows, to successive generations of synthetically related aa, during initial column-by-column code expansion. Putative cofactor/adaptor molecules normally appeared to diversify from a sibling aa tRNA; principally, an amide-amino-acid-adaptor. Primordial aa precursors, aspartate and glutamate, were identified as likely substrates for the first aminoacyl-tRNA synthetase enzymes. The pattern of aa specificity among class I and II synthetases was attributed to the imprint of code structure on synthetase evolution.

ABBREVIATIONS

A	adenine;
C	cytosine;
G	guanine;
U	uridine;
N	a standard nucleotide;
R	purine;
Y	pyrimidine;
Aa	amino acid;
Ala, A	alanine;
Arg, R	arginine;
Asn, N	asparagine;
Asp, D	aspartate;
Cit	citrulline;
Cys, C	cysteine;
f-Met	N-formyl-methionine;
Glu, E	glutamate;
Gln, Q	glutamine;
His, H	histidine;
Ile, I	isoleucine;
Leu, L	leucine;
Lys, K	lysine;
Met, M	methionine;
Orn	ornithine;
Phe, F	phenylalanine;
Pro, P	proline;
Sec	selenocysteine;
Ser, S	serine;
Thr, T	threonine;
Trp, W	tryptophan;
Tyr, Y	tyrosine;
Val, V	valine;

α-KG	α-ketoglutarate;
OAA	oxaloacetate;
3PG	3-phosphoglycerate;
Pyr	pyruvate;
Shi	shikimate;
P	probability;
SEM	standard error about mean;
tRNA-I	type I,
-II	type II,
-A	core group A,
-A'	modified core group A,
-B	group B,
-C	group C,
-D	group D,
-D'	modified core group D;
Fd	ferredoxin;
p-aaRS	protein-aminoacyl-tRNA synthetase;
r-aaRS	ribozyme-synthetase

INTRODUCTION

Since Crick's prescient conjecture that H-bonding adaptor molecules decoded template during protein synthesis, [1] tRNA has been found to participate in a range of processes beyond translation of a gene transcript. [2, 3] Synthesis of f-Met, Gln^2, Asn^2, Sec and Cys^5, and some biomolecules unrelated to amino acids (aa), involve a tRNA cofactor. [4-9] tRNA also regulates gene transcription through an interaction with the T-box in a nascent mRNA molecule. [10] In addition, pre-tRNA catalyzes self-excision of an intron in the anticodon loop. [11] tRNA participation in aa synthesis, especially among various prokaryote species, supports the suggestion that an extensive network of RNA-dependent pathways once produced aa incorporated into proteins. Beyond this, growth of these early pathways conceivably helped shape the genetic code. [12]

The proposition that growth of early aa synthesis pathways and code formation were somehow linked appears far-sighted, when viewed from the perspective of the path-distance model of code evolution. [13-15] This model equates the time-order of aa addition to the code with the number of reactions (from the citrate cycle) required for synthesis and it has been shown to unify all significant features of code structure reported over the last four decades. They include: (1) Woese clusters involving NAN column triplets assigned to polar aa residues and NUN triplets to hydrophobic residues. [16] (2) 5'-Base invariance among codons for synthetically related aa. [17] (3) Location of eight subdivided code boxes in the Standard Code. [18] (4) Allocation of intact code boxes to the six smallest aa incorporated into proteins. [17] (5) Presence of G/C enriched triplets in intact code boxes. [19] (6) Elevated coding capacity of codon mid-base. [20] (7) Dillon's codon-set subdivision rule: single codons formed by subdivision of doublets, which in turn formed from a 'set of four'.

[21] (8) NAN column triplets encode four NH_4^+ fixer aa. [13] (9) Codon mid-base, among short-path (up to 7 steps) aa, correlates with path length. [15] (10) Use of an intact code box to encode the smallest aa. [15] (11) Assignment of one codon doublet to a long-path aa and another to a short-path aa in most subdivided code boxes. [15] (12) The preferential conservation of short-path aa residues, among proteins that evolved before the last common ancestor (LCA). [14] (13) Synthesis of acidic aa residues on short paths and basic residues on long paths. [13, 14] (14) Confinement of negatively charged aa residues to a single code box. [15] (15) Fall-off in C atom oxidation number with incrreasing path length, among aa originating from citrate cycle components. [12, 15] (16) Source of Biro clusters involving allocatioon of NRN column triplets to charged residues and NYN triplets to hydrophobic residues. [15, 21] In addition, a frequency-shift model of code evolution, [22] proposing that early-comers to the code have elevated frequency in the residue sequence of an evolutionarily deep protein, displays reasonable statistical agreement ($p = 0.103$) with the path-distance model. [15] Residue frequency-shifts indicate GNN row triplets code for an earlier set of aa than ANN row triplets [23] and this also agrees with the path-distance model.

Evidence was sought in this endeavor that tRNA had a dual role as cofactors in aa synthesis and as adaptors in translation, before species diverged from the LCA. The objective being to help clarify the molecular mechanism responsible for coordinating code evolution with growth of aa synthesis pathways. By focussing this investigation on analysis of the conserved trace of LCA tRNA in extant tRNA, post-divergence base substitutions were effectively 'filtered-out'. Since they contribute around two-thirds of the variability in extant adaptor sequences, [24] mutations from this source undoubtedly impeded earlier attempts to uncover the path of pre-LCA tRNA diversification directly from extant tRNA sequences. [25-27] An approach based on eliminating post-divergence variability prior to analysis of ancestral base sequences, however, posed some challenges: (1) Information on pre-LCA tRNA evolution had to be extracted from the trace of an LCA tRNA conserved in, at most, fifty sites within an extant tRNA sequence, as almost twenty sites remained invariant throughout tRNA evolution. And, (2) LCA tRNA species for different amino acids were paralogs from a single taxonomic source, so no outgroup existed to help locate the root of their phylogenetic tree.

Different types of tRNA present in the LCA are initially identified in this chapter, following a detailed comparison of core structures and sequence identity levels. A family of synthetically related aa is then revealed to charge a particular kind of tRNA adaptor. Phylogenetic analysis of mutation distances and quaternary identity levels in the trace of LCA tRNA, conserved within extant tRNA sequences, corroborated this finding and also indicated that adaptors for same-family aa generally descended from a common ancestor. The structure of the genetic code is then interpreted from the perspective of these findings. This chapter concludes by presenting aspects of aa synthesis bearing on the mechanism of pre-LCA aa synthesis and by relating diversification of aminoacyl-tRNA synthetases (aaRS) to code evolution.

The results presented suggest that the genetic code evolved at a time when aa, synthesized from a common precursor, were attached to tRNA cofactor/adaptor molecules descended from a common ancestor. tRNA-dependent aa synthesis, integrated in this manner, was attributed with channeling assignment of nearest-neighbor codons to same-family aa.

These findings provide an explanation for the genetic code having five row-oriented coding domains.

RESULTS AND DISCUSSION

Types and Subtypes of LCA tRNA

Figure 1 shows base sequences containing the trace of LCA tRNA conserved in forty-eight extant tRNA species, obtained from the Bayreuth database. [28] These sequences strictly conserved, on average, thirty-three sites (anticodon triplet, ACC acceptor stem omitted); each conserved site contained an identical base in the consensus sequence representative of a tRNA species within each of the three domains of life. Analysis of residue sequences in pre-divergence proteins [14] previously revealed that monomers conserved across species domains exhibit 93-100 per cent identity with corresponding monomers in the LCA sequence reconstructed after establishing phylogenetic trees (see Methods, tRNA sequences). Around half (seventeen) of the conserved sites in the present survey were identical among all tRNA species, indicating they had remained invariant during tRNA evolution. Information related to tRNA diversification during code evolution therefore resided in the remaining (sixteen) non-universal conserved sites (see Methods, figure 9).

Thirty-nine type I tRNA conserved a trace of LCA aa adaptors. They separated into three subtypes, with distinct core structures (figure 1). Bases in the D stem, its surround, and variable loop of LCA tRNA species showed they had core structures recognizable from the L forms of tRNA [29] in phylogenetically deep archaea. The remaining nine adaptors were type II tRNA adaptors for Ser^4 and Leu^7.

Generic LCA tRNA-I with a group A core (figure 2) contained a D stem with 4 b.p. between segments **8-UAGCUCAG** and 21-**AGAG**$^{\mathbf{C}}{}_{\mathrm{U}}$**G**, and a 5 nt. variable loop 44-**AGGUC** (bold letters signify conserved bases). An archaeal tRNA-IA core [29] differs from this at five sites: (archaea → LCA): R9A, Y13C, R26G, N44A, and Y47U (tRNA-$IA^{Arg}{}_{3'GCA}$ contained, in addition, a C25U substitution). These differences reflect improved base resolution in the present study, with a larger number of base sequences (1100 vs. 36) and more phylogenetically diverse sources (species domains, 3 vs. 1).

Twenty tRNA-I species conserved evidence of a group A LCA ancestor (figure 1). Seventeen were charged by six Asp-derived aa [30-32]: Asn^2, Thr^6, Met^7, Ile^7, Arg^9, Lys^{10}. The remaining three ancestral tRNA-IA were charged by Ala^4, a Pyr-derived aa [33]. Five tRNA-I conserved evidence of an ancestor with a modified group A core. Pyruvate-derived aa, Val^5, acylated three tRNA-IA', and Ser^4 family members, Cys^5 and Trp^{14}, each charged a tRNA-IA'.

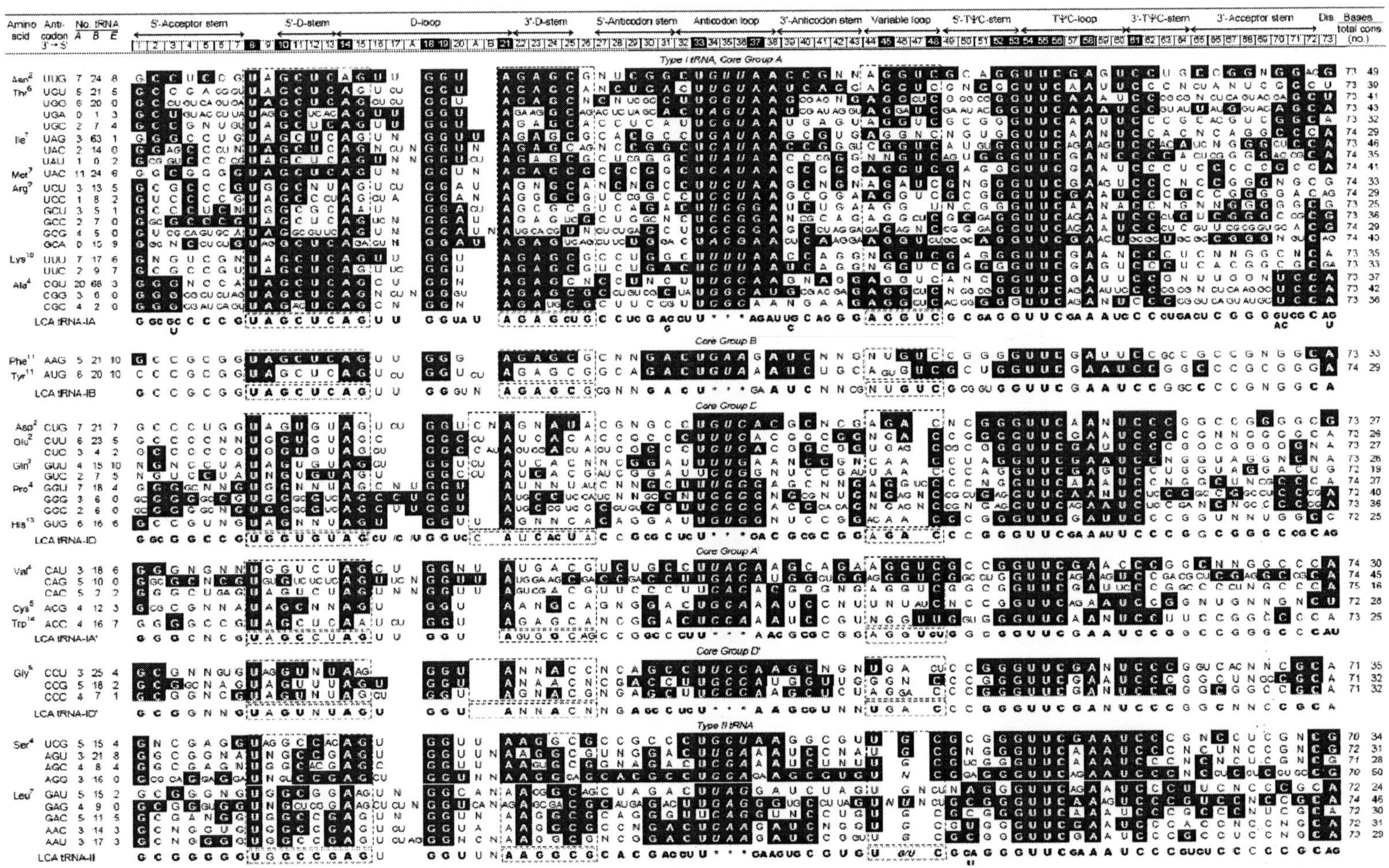

Figure 1. Trace of LCA tRNA conserved in extant tRNA base sequences. Amino acid identity, core structure type and anticodon of each tRNA species is shown. Base sequences were reconstructed from consensus sequences from an indicated number of sources in Archaea (*A*), Bacteria (*B*), and Eukarya (*E*) (Ref. 52 gives tRNA

sources and access numbers in Bayreuth database). Sites conserved across species domains in a tRNA species are highlighted in its sequence. Highlighted in the overbar are sites conserved across species domains in the consensus sequences of virtually all tRNA species. Modified bases appear in their unmodified form. N, sites with no consensus base. Dashed lines enclose D-stem and variable loop bases in the core of a tRNA L-form. The conserved trace of LCA tRNA in thirty-nine type I tRNA species charged by eighteen aa is shown. They are divided into subtypes IA, IB, and ID. Modified core groups IA' and ID' also occur. Nine type II tRNA conserved traces of LCA adaptors. A generic sequence representative of each type, and subtype, of LCA tRNA is shown; bold letters mark conserved bases. Italic letters mark a gap (sites, 45-46) in type II tRNA sequences, indicating the location of supernumerary variable-loop bases. Amino acid superscripts specify stage at addition to the genetic code in path-distance model. This survey included 1100 tRNA sequences obtained in a search of the Bayreuth database (http://www.uni-bayreuth.de/departments/biochemie/tRNA/).

Pre-divergence adaptors for Phe^{11} and Tyr^{11} had a 4 b.p. D stem and 5 nt. variable loop, similar to archaeal tRNA-IB adaptors. Their variable loop also contained a 45U (figure 1), as in an archaeal group-B core [29]. Unlike the novel G10:45G pair in tRNA-IA, a tRNA-IB has a G10:45U pair (figure 2) with a shorter inter-chain distance. Purine-purine pair, G26:44A, is ubiquitous among tRNA-IA. In an LCA tRNA-IB, however, it is not strongly conserved (figure 1, 2). Variable-loop sites 44, 45 thus distinguish ancestral aromatic aa adaptors from tRNA-IA: 44-**AG** (tRNA-IA) → 44-NU (tRNA-IB). On the other hand, archaeal tRNA-IB have a U11:24A pair [29], while LCA tRNA-IB had a C11:24G pair, as in a group IA adaptor (figure 2). Group B is viewed as a subset of group A, in archaeal tRNA. [29] A closer kinship in the LCA conforms with tRNA-IB being an off-shoot from tRNA-IA diversification (see figure 6). It also implies 45U is an identity element in the recognition of $tRNA\text{-}IB^{Phe}$ and $tRNA\text{-}IB^{Tyr}$.

Ancestral tRNA-ID had 4 b.p in their D-stem and 4 nt. variable loop, with site-47 lacking, except for Pro^4 adaptors (figure 1). A supernumerary D-loop base occurred at 20A. Two departures from an archaeal group-ID core [29], R9G and K11U (K denotes G or U), reaffirm improved resolution in the present, larger survey. At site 24, pre-divergence tRNA-ID conserved an A, or C, analogous to an archaeal group ID core. D-Stem b.p. U11:24A(C), G12:23C, and U13:22U in tRNA-D differ from the corresponding tRNA-IA and -IB b.p., C11:24G, U12:23A, C13:22G. An unusual U13:22U pair occurs in the D-stem of all tRNA-ID. An isomorph, with a U13:21A pair, shifts U22 into a mini-loop (figure 2). Triple bonds C25:G10:G45 and C25:G10:U45 in tRNA-A and -B, respectively, are replaced by a U25:G10:G45 bond in tRNA-ID. Group D tRNA-I also have a U13:A21:A46 triple, instead of the C13:G22:G46 bond in tRNA-A and -B. As noted, a strongly conserved G26:44A bond occurs in tRNA-A, while tRNA-D generally form an A26:44A bond.

Consistent with archaeal tRNA, [29] four Glu family aa (Glu^2, Gln^2, Pro^4, His^{13}) and Asp^2 account for all nine LCA group D tRNA-I (figure 1). Early $tRNA^{Gly}$ also displayed a modified group D core (figure 2). D-stem sites 10-15 were **GUNUAG** in $tRNA^{Gly}$ and **GUGUAG** in generic LCA tRNA-ID (figures 1, 2). The ancestral Gly adaptor also had a 4 nt. variable loop, 44-**U**GA-C, with a gap at site-47, and, apart from an A44U substitution, this matches an ancestral tRNA-ID variable loop, 44-AGA-C. In archaea, $tRNA\text{-}I^{Gly}{}_{CCC,3'CCG}$ and $tRNA^{Val}{}_{3'CAG}$ are designated as type IC tRNA, [29] characterized by a 10-GUCUA D-stem segment and 5 nt. variable loop, 44-NRANC.

Pre-divergence $tRNA\text{-}I^{Val}{}_{3'CAG}$ variable loop, 44-${}^{A}{}_{G}$**GGUC**, closely matches the generic LCA group-A tRNA-I variable loop, 44-**AGGUC** (figure 1). This distinguishes it from an archaeal group-C adaptor, with a 44-NRANC variable loop. The D stem of this Val adaptor was insufficiently resolved to classify. Archaeal Val isoacceptor, $tRNA\text{-}I^{Val}{}_{3'CAU}$ has a modified group A core. [29] All three pre-divergence Val adaptors, $tRNA\text{-}I^{Val}{}_{3'CAY,3'CAG}$ in this survey, had a 44-A**GGUC** variable loop and a C25:G10:G45 triple bond, consistent with being type IA related tRNA. In this connection, all LCA isoacceptors examined displayed core structure homology (figure 1, 2). This favored shared ancestry and common identity elements in isoacceptors recognized by a single aaRS. Clearly, tRNA gene recruitment occurred far less frequently during biological evolution than experimental observations would imply [29, 34], and this parallels the constrained malleability of the genetic code. [35]

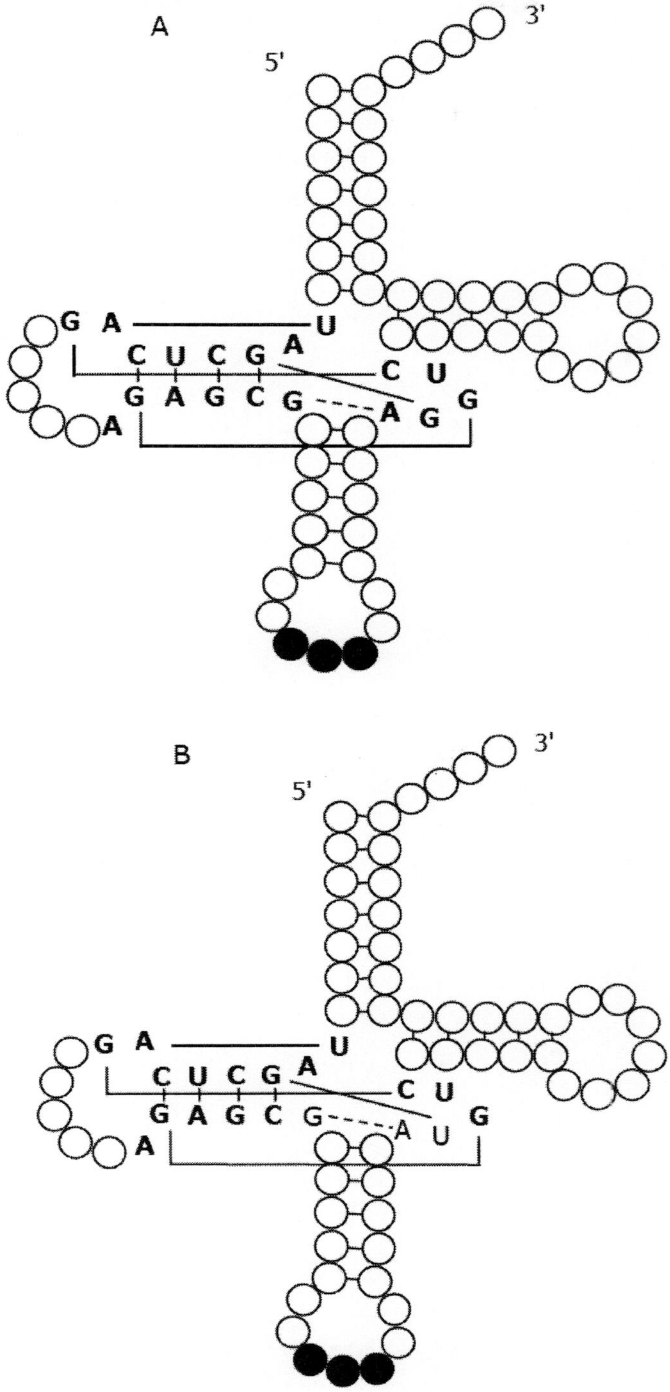

Figure 2. Continued on next page.

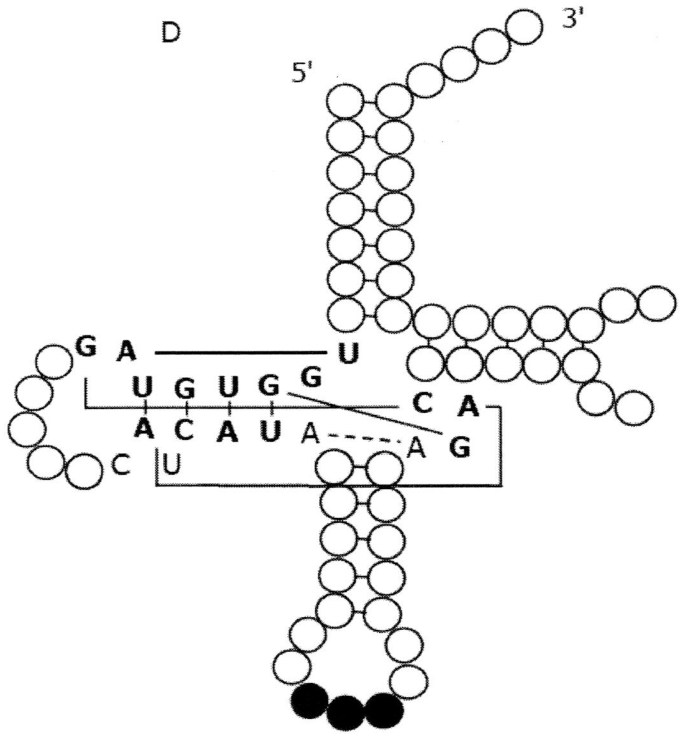

Figure 2. Subtypes of LCA type I tRNA. Cloverleaf structures of LCA tRNA with IA, IB and ID core structure are shown. Bold letters mark D stem and variable-loop bases conserved over three species domains. Lines connect base pairs and triples in tRNA L form. A dashed line depicts a non-standard purine-purine bond at 26:44. Solid circles highlight anticodon triplet. tRNA secondary structures depicted are based on generic LCA tRNA sequences in figure 1.

A modified group-A core occurred in LCA adaptors for Cys[5] and Trp[14]. They had an A-like D-stem, 10-**GC**NN**AG**/21-**A**$^{A}_{G}$N**G**CA, 5 nt. variable loop (44-NN$^{G}_{U}$**U**$^{C}_{U}$) and a C25:G10:N45 triplet. Both are Ser-derived aa. [30,32] This suggests a group A tRNA-I Ser adaptor was the source of type II tRNA adaptors.

Core Structure and Sequence Identity in LCA tRNA Species

Results presented in this section show that LCA tRNA exhibited highest sequence identity with generic tRNA possessing the same core structure group. This supports the existence of four kinds of LCA tRNA, and it corroborates phylogenetic evidence [29] that type I archaeal tRNA includes distinct subtypes.

Seven tRNA-I species displayed identity at all nineteen non-universal sites conserved jointly with generic tRNA-IA (figure 3). This corresponds to a quaternary identity, I, of 19 quarts (see Methods, Phylogenetic analysis). Average identity among all twenty tRNA-IA species with generic tRNA-IA equaled 15.73 ± 0.43 quarts (8 ± SEM) and this corresponds to 17.7 identical sites among 19 relevant sites. Mean identity versus generic tRNA-IB and -ID was significantly lower: 11.82 ± 0.38 and 2.19 ± 0.20 quarts, respectively, or around 16 and 8 identical sites over 19 sites. Ranking these mean identities revealed $)_A$(A) > $)_A$(B) > $)_A$(D), at $p < 1x10^{-3}$, in a least significant difference analysis [36], where $)_A$(A), $)_A$(B) and $)_A$(D) refer to the mean identity of tRNA-IA species with an indicated generic tRNA-I subtype. Arginine[9] isoacceptors, tRNA-IA^{Arg} $_{UCU,3'GCU}$ showed least identity with generic tRNA-IA; $)_{Arg}$(A) equaled 6.88 quarts (identity,13/19 sites). This still exceeded their identity with generic tRNA-IB and -ID; $)_{Arg}$(B) and $)_{Arg}$(D) equaled 4.56 quarts (identity, 11/19 sites) and 3.90 quarts (identity, 11/20 sites), respectively.

Highest identity between tRNA-IB species and generic tRNA, likewise, coincided with core group homology (figure 3). $)_B$(B) was 16.08 quarts (identity, 18/19 sites), while $)_B$(A) equaled 14.89 quarts (identity, 17.5/19 sites) and $)_B$(D) equaled only 2.10 quarts (identity, 6.5/20 sites). Least significant difference analysis ranked these mean identities as: $)_B$(B) ≈ $)_B$(A) > $)_B$(D) at $p < 1x10^{-3}$.

Among tRNA-ID species, $)_D$(D) equaled 9.63 ± 1.24 quarts (ca.15/19 sites). Mean identity with heterologous generic tRNA was again lower: $)_D$(A) and $)_D$(B) were merely 2.55 ± 0.28 and 1.95 ± 0.16 quarts (identity, ca. 8/19 and 7/19 sites), respectively. These means ranked, $)_D$(D) > $)_D$(A) ≈ $)_D$(B), at $p < 1x10^{-3}$. tRNA^{His} displayed least homology with generic tRNA-ID; I_{His}(D) = 3.62 quarts (identity, 10/19 sites), equaling I_{His}(A) and exceeding I_{His}(B).

tRNA containing a modified group IA core structure showed highest mean identity with generic tRNA-IA; $)_{A'}$(A) equaled 7.08 ± 1.09 quarts (identity, ca. 13/19). While $)_{A'}$(B) and $)_{A'}$(D) were, respectively, 5.80 ± 0.83 and 4.14 ± 1.13 quarts (identity, 12/19, ca. 10/19). Least significant difference analysis ranked these mean identities as: $)_{A'}$(A) > $)_{A'}$(D) at $p < 1x10^{-2}$, with $)_{A'}$(B) not significantly different from either $)_{A'}$(A), on the highside, or $)_{A'}$(D), on the lowside. Modified group D tRNA had mean quaternary identity versus generic tRNA-ID, -IA and -IB equal to 4.61 ± 0.58 (identity, 11/19 sites), 4.03 ± 0.84 (identity, ca. 10/19 sites), and 2.88 ± 0.42 (identity, 9/19 sites), respectively. They were ranked, $)_{D'}$(D) > $)_{D'}$(A) ≈ $)_{D'}$(B) at $p < 5x10^{-2}$.

Type-II tRNA species displayed highest identity (supernumerary variable loop, anticodon and ACC acceptor stem bases omitted) at non-universal sites jointly conserved with a generic tRNA-IA and -IB. $)_{II}$(A) and $)_{II}$(B) both equaled 4.47 ± 0.20 quarts (identity, 11/19 sites). Whereas, $)_{II}$(D) was only 1.68 ± 0.11 (identity, ca. 7/19 sites). Sequential mean difference analysis ranked these means as, $)_{II}$(A) = $)_{II}$(B) > $)_{II}$(D), at $p < 1x10^{-3}$.

Each type/subtype of LCA tRNA identified in the previous section exhibits a characteristic pattern of base interactions in its D stem and variable loop. Sequence identity levels presented in this section demonstrate that core group homology implies tRNA sequence homology. This corroborates an earlier finding that phylogenetic kinship in an archaeal tRNA tree implies core structure group homology. [29]

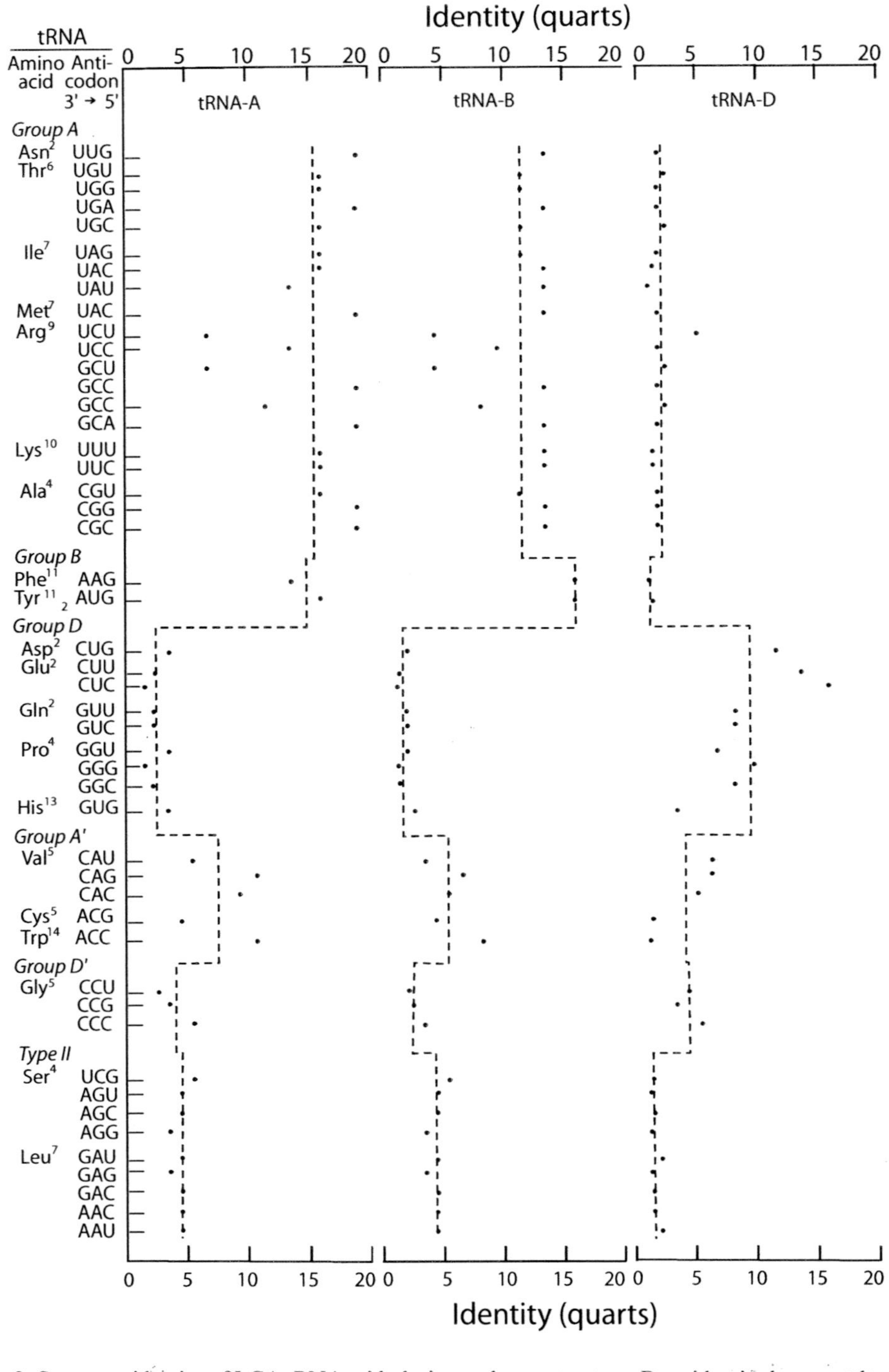

Figure 3. Sequence identity of LCA tRNA with designated core structure. Base identity between the conserved trace of LCA tRNA and generic sequence of LCA tRNA-I subtypes is presented. Dashed vertical lines show mean trace identity, in quaternary units (quarts), at jointly conserved, non-universal sites (anticodon and supernumerary variable-loop bases omitted) among tRNA possessing a designated core structure. Significantly higher mean identity occurred between LCA tRNA and generic tRNA-I containing the same core group. tRNA core structures were determined from base sequences in figure 1.

tRNA containing a modified group IA core structure showed highest mean identity with generic tRNA-IA;)$_{A'}$(A) equaled 7.08 ± 1.09 quarts (identity, ca. 13/19). While)$_{A'}$(B) and)$_{A'}$(D) were, respectively, 5.80 ± 0.83 and 4.14 ± 1.13 quarts (identity, 12/19, ca. 10/19). Least significant difference analysis ranked these mean identities as:)$_{A'}$(A) >)$_{A'}$(D) at $p < 1x10^{-2}$, with)$_{A'}$(B) not significantly different from either)$_{A'}$(A), on the highside, or)$_{A'}$(D), on the lowside. Modified group D tRNA had mean quaternary identity versus generic tRNA-ID, -IA and -IB equal to 4.61 ± 0.58 (identity, 11/19 sites), 4.03 ± 0.84 (identity, ca. 10/19 sites), and 2.88 ± 0.42 (identity, 9/19 sites), respectively. They were ranked,)$_{D'}$(D) >)$_{D'}$(A) ≈)$_{D'}$(B) at $p < 5x10^{-2}$.

Type-II tRNA species displayed highest identity (supernumerary variable loop, anticodon and ACC acceptor stem bases omitted) at non-universal sites jointly conserved with a generic tRNA-IA and -IB.)$_{II}$(A) and)$_{II}$(B) both equaled 4.47 ± 0.20 quarts (identity, 11/19 sites). Whereas,)$_{II}$(D) was only 1.68 ± 0.11 (identity, ca. 7/19 sites). Sequential mean difference analysis ranked these means as,)$_{II}$(A) =)$_{II}$(B) >)$_{II}$(D), at $p < 1x10^{-3}$.

Each type/subtype of LCA tRNA identified in the previous section exhibits a characteristic pattern of base interactions in its D stem and variable loop. Sequence identity levels presented in this section demonstrate that core group homology implies tRNA sequence homology. This corroborates an earlier finding that phylogenetic kinship in an archaeal tRNA tree implies core structure group homology. [29]

LCA Adaptors for Sibling Amino Acids Display Core Structure Homology

All Asp-derived aa charged tRNA-IA adaptors in the LCA (table 1). Both Shi-derived aa used tRNA-IB adaptors. While Glu^2 and all three Glu-derived aa charged tRNA-ID adaptors.

Two Pyr-derived aa and two Ser family siblings acylated type IA related tRNA adaptors. The results show synthetically related aa preferentially charged homologous tRNA in the pre-divergence era (see Methods, LCA tRNA types and subtypes).

tRNA core structure distributions in the Asp and Glu family (precursor + siblings) were at, or near, statistically significant levels of homogeneity: $p = 5.5x10^{-2}$ (Asp family) and $3.9x10^{-3}$ (Glu family). tRNA homology occurred in the small Shi family; $p = 1x10^{-2}$. Combining tRNA distribution probabilities for all five families [37] revealed a close correlation existed between aa source and LCA tRNA core structure; $p = 1.5x10^{-3}$. Parenthetically, a statistically significant combined probability, $p = 3.38x10^{-2}$, resulted even when Phe and Tyr adaptors were classified as type-IA tRNA.

Among adaptors for precursor aa, Asp^2, Glu^2, Ser^4, only the Glu-adaptor, tRNA-ID^{Glu}, displayed within-family homology: a within-family precursor tRNA homogeneity of 33 per cent. By comparison, product aa, from five families, charged fifteen LCA tRNA exhibiting within-family homogeneity (table 1), Gly^5 and Leu^7 adaptors being the only exceptions: a within-family precursor tRNA homogeneity of 88.2 per cent. Conspicuously lower precursor-tRNA homogeneity is at variance with the precursor-product hypothesis [38], whereby a product aa purportedly entered the code by capturing a precursor-adaptor. It appears from the tRNA distribution in table 1 that a product aa entered the code more frequently by acquiring a variant sibling adaptor.

Results presented in the following section corroborate the significance of tRNA descent from a sibling aa adaptor. They also provide evidence that the pre-LCA growth of aa synthesis pathways, tRNA diversification and code evolution were coordinated.

Table 1. Distribution of pre-divergence tRNA types and subtypes among amino acid families[1]

Amino acid	tRNA I A	tRNA I B	tRNA I D	tRNA II	p^2	Amino acid	tRNA I A	tRNA I B	tRNA I D	tRNA II	p
		Aspartate family						*Pyruvate family*			
Asp[2]			+			Ala[4]	+				
Asn[2]	+					Val[5]	+				
Thr[4]	+					Leu[7]				+	
Met[7]	+					Distn.	2	0	0	1	0.375
Ile[7]	+										
Arg[9]	+										
Lys[10]	+										
Distn.	6	0	1	0	$5.5\text{x}10^{-2}$						
		Glutamate family						*Serine family*			
Glu[2]			+			Ser[4]				+	
Gln[2]			+			Cys[5]	+				
Pro[4]			+			Gly[5]			+		
His[13]			+			Trp[14]	+				
Distn.	0	0	4	0	$3.9\text{x}10^{-3}$	Distn.	2	0	1	1	0.375
								Shikimate family			
						Phe[11]		1			
						Tyr[11]		1			
						Distn.	0	2	0	0	$1.0\text{x}10^{-2}$

Global distribution: $\chi^2_{10} = 30.04$, $p = 8.45\text{x}10^{-4}$

[1] Amino acid family members derive from a designated precursor. [30-33] Superscripts indicate the estimated stage of code evolution aa entered the genetic code, in the path-distance model. tRNA types and subtypes correspond to those in figure 1 for pre-divergence adaptors.

[2] *p*, probability the principal tRNA core structure, within each family, arose chance. Global probability is the combined family probability. [37]

Channeled Diversification of tRNA Species in the Pre-LCA Era

Neighbor-joining calculations [39] with Jukes-Cantor distances [40] at non-universal sites shared between the conserved trace of LCA tRNA produced the dendogram in figure 4 (see Methods, Phylogenetic analysis). The tree reconstructs tRNA diversification in the pre-LCA era from vestiges of ancestral tRNA conserved within populations of extant tRNA base sequences.

Construction of this tree follows the procedure illustrated in uncovering a 23-residue Fd antecedent (Pro-Fd-5) from the vestige of a tandem duplication in the 55-residue sequence of LCA ferredoxin (Fd). [14, 41] Statistical evaluation of the tree topology obtained on analysis of mutation distances between homologues of Fd from anaerobe Clostridium pasteurianum relied on demonstrating a fall-off in internal identity with increasing distance from the tree root. Rather than assess the tree from a set of nodal bootstrap probabilities, a single probability thus indicated whether it was globally ordered. In the event that early tRNA diversification and growth of aa synthesis pathways were linked, as suggested by results in the previous section, the path-distance model would favor early (short-path) aa having early tRNA adaptors. tRNA ancestral to extant adaptors with short-path aa accordingly should cluster near the tree root. In principle, this provides a means to locate the root of the LCA tRNA tree (figure 4); since these tRNA are from a single taxonomic source, use of outgroups was not an option. Furthermore, aa code age could be anticipated to increase with tRNA distance from the tree root in a valid topology.

Figure 5 shows that aa code age increased linearly, at 1.57 stages per unit distance, in the tree depicting pre-LCA tRNA diversification (figure 4). Tree topology supports a positive increase in aa code age with tRNA distance at a probability, $p(\beta > 0) = 0.948$ ($t(df = 30) = 1.671$), where β is the regression equation slope. [36] This increase occurred over stages 2-7 of code evolution, corresponding to initiation and expansion phases of code evolution in the path-distance model. [13-15] The correlation does not extend to post-expansion phase, when Arg^9, Lys^{10}, Phe^{11}, Tyr^{11}, His^{13}, Trp^{14} were added to the code. This is anticipated, as they apparently captured short-path aa adaptors, following subdivision of error-prone code boxes (sets of four) into doublets. [13,15,18] $Serine^4$ adaptors were also non-conforming, and these type II tRNA evidently replaced the type I tRNA originally charged by this early expansion phase aa (see Types and subtypes of LCA tRNA, Identity and source of LCA tRNA).

LCA tRNA tree topology (figure 4) incorporated a second global feature: aa synthesized from different precursors differ in tree stacking rank. Differences in aa family stacking rank arose by chance with a probability, $p = 2.05 \times 10^{-2}$ ($\chi^2(df = 4) = 11.61$), in a Kruskal-Wallis analysis. [42] Inspection of the LCA tRNA tree reveals that adaptors for same-family aa cluster within distinct 'monophyletic' tree regions, excluding both Shi family adaptors - $tRNA\text{-}IB^{Phe}{}_{3'AAG}$, $tRNA\text{-}IB^{Tyr}{}_{3'AUG}$ (figure 4). At the tree base, region-1 contains eight tRNA-ID species and $tRNA\text{-}II^{Leu}{}_{3'GAU}$; as this type II tRNA is in a type I tRNA region, gene recruitment [29, 34] can be apparenty excluded. Seven are Glu family adaptors; $tRNA\text{-}ID^{Asp}$ being the only non-Glu family type-ID adaptor. Anticodons of five region-1 tRNA contain a 35U and six had a 36G (anticodon mid- and 3'-base). Three aa (Asp^2, Glu^2, Gln^2) with region-1 tRNA adaptors were NH_4^+ fixers. They form on 1-2 step paths and are first-generation aa in the path-distance model of code evolution. [13, 15] Amino acid path-distances thus place region-1 at the root of the tRNA tree. Branch lengths in region-1 and -2 are scaled to one-tenth their original length, suggesting tRNA initially changed more rapidly than the repertoire of aa.

Pyruvate-derived aa charged six (6/8) tRNA in region-2. All were type IA tRNA. Isoleucine^{7} also charged a region-2 type IA tRNA, raising the possibility of gene recruitment. tRNA-ID$'^{Gly}{}_{3'CCG}$ was the only non-type IA adaptor in region-2. Seven region-2 tRNA had a 36C. Region-3 contained eighteen tRNA. Fourteen were adaptors for Asp-derived aa and they charged type IA tRNA. Twelve had a 36U. Among thirteen tRNA in region-4, Ser family aa acylated eight. Seven type II tRNA branched in region-4. Five region-4 tRNA contained a 36A and six a 35C. Adaptors for Shi derived aa, Phe^{11}, Tyr^{11}, branched in region-3 (figure 4). Consistent with this, identity levels indicate their type IB adaptors descended from a tRNA-IA (see table 2); they form a fifth cluster-region in archaeal (Haloferax volcanii) tRNA (unpublished result).

The pattern of pre-LCA tRNA diversification apparent in figure 4 occurs, in muted form, in the tree constructed from sequences (possessing post-divergence variability) of thirty-six archaeal tRNA species by the Fitch-Margoliash method (range of nodal bootstrap probabilities, $p = 0.36\text{-}1.0$). [29] As in figure 4, region-1 cluster had adaptors for Glu^{2}, Gln^{2}, Pro^{4}, His^{13}, Asp^{2}, and region-2 had Ala^{4}, Val^{5} adaptors. Region-3 contained Thr^{6}, Met^{7}, Ile^{7}, Lys^{10} adaptors, while Ph^{11} and Tyr^{11} adaptors separated into a distinct cluster (region-5). Adaptors for Ser^{4}, Cys^{5}, Trp^{14} and Leu^{7} (region-4) were omitted from this analysis. Adaptors for thirteen aa (13/16) thus cluster in archaeal tree regions 1-3 and 5, corresponding to an 81 per cent cluster-rate. This compares favorably with the 85 per cent (17/20) rate evident in the LCA tree (figure 4). Consistent with this agreement, archaeal tRNA were found to have evolved only slowly. [27] tRNA adaptors for synthetically related aa cluster more in evolutionarily deep sources, in accord with tRNA-dependent aa synthesis in the pre-LCA era (unpublished result).

Same-family aa preferentially charged phylogenetically-related tRNA, in the pre-LCA era, according to present results. Combining probabilities for tRNA aa identity distribution, within each region, revealed significant clustering in the tRNA tree; $p = 1.36\text{x}10^{-8}$, χ^2 (df = 10) = 56.90 (see Appendix A: Distribution of amino acid identity, core group and anticodon bases in LCA tRNA tree). tRNA core structure also showed significant clustering within tree regions, on a global scale; $p = 1.21\text{x}10^{-6}$, χ^2 (df = 10) = 46.42. 3'-Anticodon base identity clusters were equally significant; $p = 1.04\text{x}10^{-6}$, χ^2 (df = 10) = 46.77. In contrast, mid-base identity clusters were comparatively weak, $p = 4.47\text{x}10^{-2}$, χ^2 (df = 10) = 18.67.

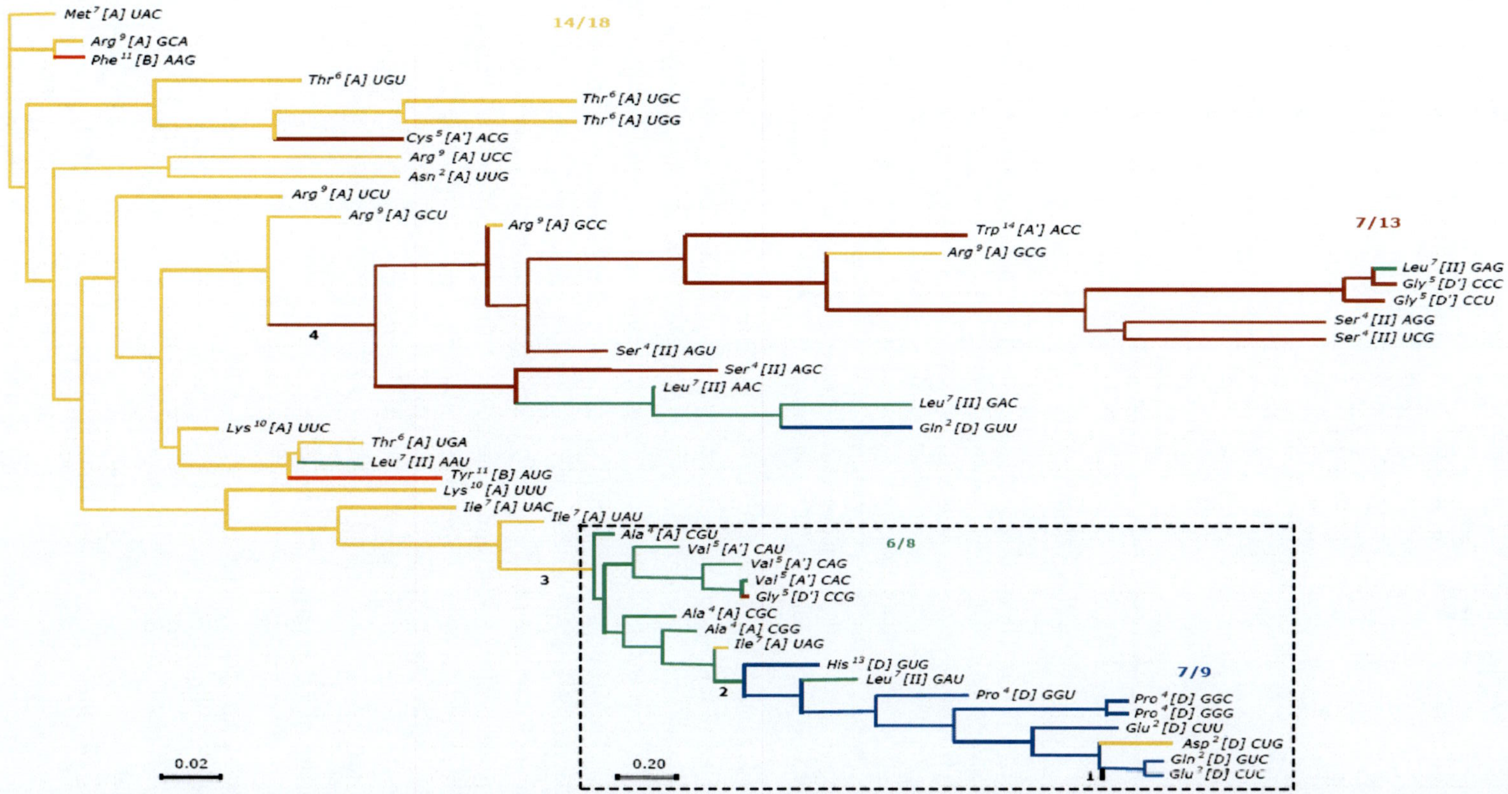

Figure 4. Dendogram depicting tRNA diversification in the pre-LCA era. The tree contains forty-eight tRNA species and was constructed from Jukes-Cantor distances at jointly conserved, non-universal sites in the conserved trace of LCA tRNA. Amino acid identity, core type and anticodon bases (3' → 5') of tRNA are indicated. The tree root is located at branch 1, which bifurcates into a cluster of tRNA charged by NH_4^+-fixer aa (Asp^2, Glu^2, Gln^2), identified as first-generation aa in the path-distance model of code evolution. [13, 15] Same-family aa preferentially charged tRNA branching within tree regions rooted at branches 1 to 4. Orange branches indicate Asp family tRNA adaptors; blue, Glu family; green, Pyr family; brown, Ser family; and, red, Shi family. Number of tRNA charged by synthetically related aa to total number is given for each tree region. Branch lengths in region-1 and -2 are reduced 10-fold. This tree was constructed with sequences in figure 1 by the neighbor-joining method. [39]

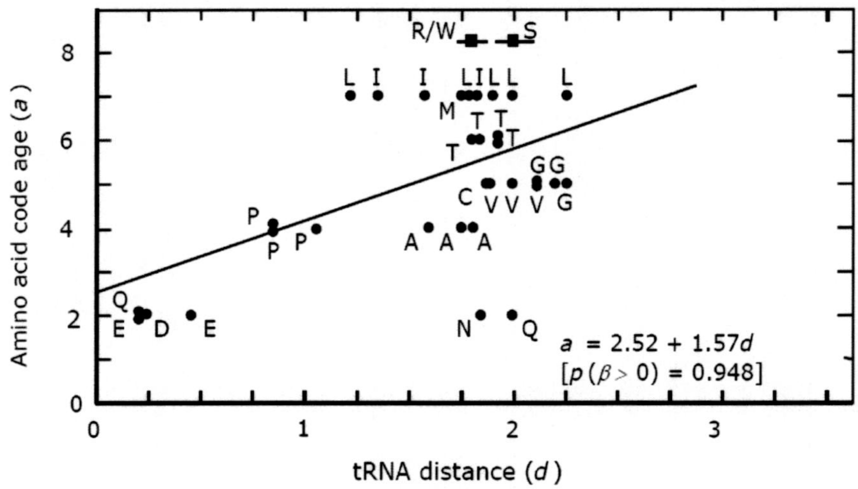

Figure 5. Correlation between aa code age and adaptor distance in the LCA tRNA tree. Amino acid code age increases linearly with adaptor distance from the tree root, between stage 2 and 7 of code evolution. The regression equation slope, β, exceeds zero with a probability, $p(\beta > 0) = 0.948$. This indicates the inferred pre-LCA tRNA phylogeny supports the path-distance model of code evolution, based on coordinated growth of aa synthesis pathways and tRNA diversity. Dots show distance of tRNA adaptors for aa designated by single letter abbreviations (see text). Squares show mean distance, with standard error bars, of indicated tRNA. R/W, tRNA-I charged by post-expansion phase (stage 9-14) additions to code: Arg^9, Lys^{10}, Phe^{11}, Tyr^{11}, His^{13}, Trp^{14}. S, type II tRNA Ser^4 adaptors. Distances are from LCA tRNA tree in figure 4.

Acquisition of phylogenetically related tRNA, by synthetically related aa, links pre-LCA tRNA diversification with growth of aa synthesis pathways during code evolution. This provides compelling evidence that tRNA were cofactors in pre-LCA aa synthesis pathways. Adaptors for related aa also displayed elevated core structure homology and 3'-anticodon base identity. Elevated 3'-anticodon base identity in same-family tRNA adaptors (figure 4) is implied by 5'-base invariance observed among codons for same-family aa. [16] Code expansion in a column-by-column pattern, through mid-base substitutions, conforms with weak clustering for anticodon mid-base and invariance of the codon 5'-base among same-family aa. [13,15] Present results point to the code containing domains of contiguous codons read by tRNA adaptors, displaying core structure homology, for same-family aa.

Identity and Source of LCA tRNA Species

The source of pre-LCA tRNA species was sought in this section from quaternary identity levels in the conserved trace of ancestral tRNA. In addition to elevated trace identity, time-order based on path-distance implies that an aa attached to the antecedent of a pre-LCA tRNA would have a shorter path. [13, 15] Metabolic source of the acylating aa (table 1), adaptor core structure (figure 1, 2), and location within the LCA tRNA tree (figure 4) were also considered in forming adaptor/ancestor pairs. Among tRNA isoacceptors, those with a 34U were chosen for this study when possible, as the path-distance model suggests U occupied the 5'-anticodon site in the initial stages of code evolution. [13, 15, 18] The source of nineteen (19/23) pre-LCA tRNA species has been reported.

Traces of LCA tRNA shared from 0 to 20 non-universal sites, conserved across species domains. Their intrinsic identity varied from 0 (Leu^7, 3'GAU → Arg^9, 3'GCU, anticodon triplets) to 100 (Asn^7, 3'UUG → Lys, UUU) per cent. While quaternary identity levels ranged from 0 to 16 quarts (table 2). Thirteen (13/19) adaptor/ancestor pairs showed same-family aa identity, two being charged by the same aa (isoacceptor pairs). Identity values for these pairs appear within triangles in table 2. Sixty-eight per cent (13/19) of adaptor/ancestor pairs identified thus involved same-family aa, despite same-family aa pairs representing only 15 per cent (37/253) of the number of pairwise combinations. Their frequency is statistically significant in a contingency table, with χ^2 corrected for continuity; $\chi^2(df = 1) = 9.21$, and $p = 2.4x10^{-3}$. tRNA in twelve same-family aa pairs also clustered together in the LCA tRNA tree (figure 4). Present results thus corroborate the earlier finding that tRNA charged by same-family aa generally descended from a common ancestor (See Channeled diversification of tRNA species in the pre-LCA era).

Serine^4 → Cys^5 (trace identity, 8.5 quarts) was the only non-conforming pair, as tRNA-II^{Ser} $_{3'AGU}$ and tRNA-IA$^{'Cys}$$_{3'ACG}$ branched in region-4 and -3, respectively. Ten adaptor/ancestor pairs also displayed core structure homology. Three heterologous pairs involved adaptors for Ser-derived aa, tRNA-IA$^{'Cys}$$_{3'ACG}$, tRNA-IA$^{'Trp}$$_{3'ACC}$ and tRNA-ID$^{'Gly}$$_{3'CCU}$ indicative of the putative type I → II tRNA transition by the Ser^4 adaptor. Elevated tRNA-II^{Ser} $_{3'AGU}$ identity with tRNA-IA^{Asn}$_{3'UUG}$ (9.2 quarts, table 2) suggests Ser^4 once charged an Asn-like, type I tRNA cognate with NCN column codons, UCN. The transition to a type II tRNA opened the way for relocation of anticodon-loop identity elements, to an enlarged variable loop, and allow assignment of multiple codon sets (NCN and NGN column triplets) to Ser^4 during code expansion from the NH_4^+ Fixers Code. [13, 15] Adaptor identity in the Tyr^{11} → Trp^{14} pair may be noted to slightly exceed that in the Ser^4 → Trp^{14} (aa precursor/product) pair: 8 vs. 7 quarts (table 2). However, tRNA^{Ser} $_{3'AGU}$ branched in tree region-4, with tRNA-IA$^{'Trp}$ $_{3'ACC}$, while tRNA-IB^{Tyr}$_{3'AUG}$ did not (figure 4), favoring the Trp-precursor adaptor as the source. In addition, the Trp-adaptor showed core structure homology with the adaptor for sibling aa, Cys^5, and putative ancestral Ser-adaptor.

tRNA-II^{Leu} $_{3'AAU}$ shows kinship with tRNA-II^{Ser} $_{3'AGU}$ (trace identity, 5.7 quarts; table 2). Nearest-neighbor codons in the UNN row (UCN, UUR) also encode Ser^4 and Leu^7. Most Leu isoacceptors, moreover, branch in tree region-4 with Ser adaptors (figure 4). These findings point to tRNA-II^{Leu} $_{3'AAU}$ descent from a variant tRNA-II^{Ser} $_{3'AGU}$ (table 2). Because the synthesis of Ser^4 and Leu^7 originate at 3PG and Pyr, [30, 31] respectively, this path of descent appears to uncouple aa pathway growth and tRNA diversification, in a departure from to earlier findings (see Channeled diversification of tRNA species in the pre-LCA era). Alternatively, an early aa family, containing both Ser^4 and Leu^7, could have fragmented following pathway rearrangement. This development plausibly accompanied the protein takeover of aa synthesis pathways, some time after appearance of the type II tRNA Leu-adaptor. With evidence that a tRNA^{Asn}-like adaptor was ancestral to the Ser-adaptor (table 2), the Asp family once conceivably included Ser family aa. Leucine^7 sibling, Ala^4, likewise charged a tRNA^{Asn}-like adaptor (trace identity, 11 quarts with pre-Lys^{10} adaptor, table 2). Inclusion of both Ser and Pyr aa families in a pre-LCA Asp family could

Table 2. Quaternary identity between conserved trace of LCA tRNA sequences[1]

tRNA		Glu				Pyr				Asp								Ser					Shi	
aa	anti-codon 3'→5'	Glu^{2} CUU	Gln^{2} GUU	Pro^{4} GGU	His^{13} GUG	Ala^{4} CGU	Val^{5} CAU	Leu^{7} AAU	Leu^{7} GAU	Asp^{2} CUG	Asn^{2} UUG	Thr^{6} UGU	Met^{7} UAC	Ile^{7} UAU	Arg^{9} UCU	Arg^{9} GCU	Lys^{10} UUU	Ser^{4} AGU	Ser^{4} UCG	Cys^{5} ACG	Gly^{5} CCU	Trp^{14} ACC	Phe^{11} AAG	Tyr^{11} AUG
Glu^{2}	CUU		2.0	1.1	2.0	1.1	1.4	1.0	1.8	**7.0**	1.3	1.3	0.96	2.7	1.8	0.71	1.5	2.2	0.62	5.7	3.0	1.1	0.62	3.0
Gln^{2}	GUU	2/2		**4.0**	1.4	2.5	4.0	2.0	1.0	3.0	3.0	1.3	3.0	**1.4**	3.0	2.0	*4.0*	6.0	5.0	5.0	3.2	4.0	3.0	6.0
Pro^{4}	GGU	2/4	4/4		2.0	3.0	1.0	0.71	3.0	1.0	**1.1**	1.8	1.5	0.62	2.2	0.62	1.0	1.4	2.2	2.0	1.5	3.0	0.71	1.4
His^{13}	GUG	2/2	2/3	2/2		5.0	1.42	1.42	4.0	1.0	2.2	2.0	0.71	3.0	0.71	2.0	3.0	2.5	2.5	3.0	3.0	0.71	0.71	1.4
Ala^{4}	CGU	2/4	4/6	4/5	5/5		**5.0**	2.5	2.5	0.67	3.7	7.4	8.3	7.0	5.0	3.0	*11.0*	4.2	2.0	4.8	1.4	2.2	12.0	5.0
Val^{5}	CAU	2/3	4/4	**1/1**	2/3	5/5		1.8	1.8	0.62	3.6	2.7	7.0	7.0	6.0	**1.0**	6.0	3.9	4.8	1.4	3.9	3.0	3.2	4.0
Leu^{7}	AAU	1/1	2/2	1/2	2/3	4/6	3/5		1.8	0.62	2.3	4.8	4.8	4.8	7.0	4.8	4.8	**5.7**	4.8	3.9	3.4	5.0	7.4	8.0
	GAU	3/5	1/1	3/3	1/1	4/6	1/1	3/5		1.3	**1.8**	3.9	2.1	2.5	3.0	0	2.2	2.2	**1.4**	1.8	1.8	3.0	3.0	2.0
Asp^{2}	CUG	7/7	3/3	**1/1**	1/1	1/5	1/3	1/4	3/7		**1.5**	1.1	0.84	**1.5**	1.3	**1.4**	0.84	1.8	1.8	0.84	5.7	2.2	0.84	1.8
Asn^{2}	UUG	3/7	6/10	3/8	3/4	9/16	7/11	6/12	4/8	4/9		10.7	**15.1**	7.4	10.0	3.9	***16.0***	**9.2**	5.7	5.7	5.3	5.7	9.9	6.5
Thr^{6}	UGU	3/7	6/10	3/8	3/4	10/12	5/8	7/9	5/6	3/8	16/20		14.0	7.5	10.0	3.0	*13.0*	5.7	2.1	7.0	4.6	5.0	15.0	7.0
Met^{7}	UAC	2/5	3/3	3/5	1/2	11/13	7/7	7/9	4/7	1/7	17/18	14/14		9.0	**13.0**	4.0	14.0	5.7	3.9	4.8	1.8	4.0	**16.0**	9.0
Ile^{7}	UAU	5/8	2/3	1/3	4/5	7/7	7/7	6/7	4/6	4/9	10/12	9/10	9/9		*9.0*	4.8	***10.0***	4.8	4.2	2.1	6.6	3.0	10.0	4.8
Arg^{9}	UCU	3/5	3/3	3/4	1/2	5/5	6/6	7/7	3/3	3/7	10/10	10/10	13/13	9/9		**7.0**	10.0	9.0	7.0	6.0	4.0	7.0	9.0	7.0
	GCU	1/2	2/2	1/3	2/2	3/3	1/1	6/7	0	2/3	5/6	4/5	4/4	6/7	7/7		5.0	**6.0**	4.0	5.0	**4.8**	4.0	6.0	4.0
Lys^{10}	UUU	1/4	4/4	1/1	4/5	11/11	6/6	7/9	3/4	2/7	16/16	13/13	14/14	10/10	10/10	5/5		8.3	4.2	8.0	3.4	6.0	15.0	7.0
Ser^{4}	AGU	3/6	6/6	2/3	4/6	7/10	5/6	7/8	3/4	4/8	12/14	8/10	9/12	7/9	9/9	6/6	11/13		**13.0**	**8.5**	5.7	**7.0**	6.5	7.5
	UCG	1/4	5/5	3/4	4/6	5/10	6/7	6/7	2/3	3/5	9/12	4/7	8/13	7/10	7/7	4/4	7/10	13/13		4.8	**8.0**	5.0	3.0	5.7
Cys^{5}	ACG	1/4	5/5	2/2	4/5	6/7	2/3	5/6	3/5	2/7	7/8	7/7	6/7	4/7	6/6	5/5	8/8	10/11	6/7		3.4	4.0	6.6	4.8
Gly^{5}	CCU	7/8	5/7	3/6	3/3	**4/10**	5/6	6/9	3/5	7/8	10/15	9/14	4/8	8/9	6/8	6/7	6/9	8/10	8/8	6/9		6.0	2.6	4.0
Trp^{14}	ACC	3/3	4/4	4/5	1/2	3/4	3/3	5/5	3/3	3/4	8/10	5/5	6/8	4/5	7/7	4/4	6/6	7/7	5/5	4/4	6/6		5.0	8.0
Phe^{11}	AAG	1/4	3/3	1/2	3/3	12/12	5/7	10/12	4/5	1/7	14/17	15/15	16/16	10/10	9/9	6/6	15/15	10/13	6/10	8/9	6/11	5/5		**10.0**
Tyr^{11}	AUG	3/3	6/6	2/3	2/4	5/5	6/8	8/8	2/2	3/5	10/13	7/7	9/9	6/7	7/7	4/4	7/7	9/10	7/8	6/7	6/8	8/8	10/10	

[1] Upper-right half of table gives identity in quaternary units (see text) among LCA tRNA sequences (figure 1). Lower-left half, number of identical sites over total number of jointly conserved, non-universal tRNA sites. Identities for tRNA charged by same-family aa are enclosed within triangles. Bold identity values mark a tRNA and its putative ancestor. Superscripts on aa give their 'code age' in the path-distance model. Italics indicate Asn tRNA was likely ancestor, prior to subdivision of code box AAN (read initially with anticodon triplet, UUU) and capture of AAR by a Lys adaptor.

account for the kinship observed between Ser- and Leu-adaptors within the framework of coordinated tRNA diversification and growth of amino acid synthesis pathways (see Channeled diversification of tRNA species in the pre-LCA era). According to this, Pyr and Ser family aa synthesis pathways respectively included central metabolism segments, OAA $\rightarrow$ Pyr and OAA $\rightarrow$ 3PG, prior to the protein-takeover.

Consistent with an enlarged pre-LCA Asp family, adaptors for fourteen aa were traced to a primordial Asn adaptor, tRNA-IA^{Asn} $_{UUU}$. They cluster in the larger of two deeply branching domains in the LCA tRNA tree in figure 6. Nodes in this tree join an LCA tRNA with its putative ancestor (figure 6a), based on quaternary identity levels (table 2). Branch lengths (ordinal value) correspond to the attached aa code age, estimated from path-distance. Progression along the y-axis thus depicts events at increasingly later stages of code evolution. [13-15] The smaller, earlier domain contained five tRNA-ID charged by Glu family aa and Asp^2. Aspartate^2 and Glu^2 are precursors, in biosynthesis pathways, to half (nine) aa. [13,15,30-33] Aspartate- and Glu-adaptors were not, however, the apparent source of product aa adaptors (table 2). This finding is counter to the 'precursor-product' hypothesis, wherein an incoming aa captured a precursor-adaptor on entering the code, [13,15,39]. Present findings instead implicate precursor misacylation of a variant sibling adaptor, generally a tRNA^{Asn}-like or tRNA^{Gln}-like adaptor, as the mechanism whereby a new aa entered the code (figure 6a).

At the tree root is a tRNA-ID. It was uncovered on backtracking from the NH_4^+ Fixers Code [13, 15]. The primordial tRNA adaptor appears to have been a pre-informational 'universal' tRNA adaptor charged by an Asp^2, Glu^2, Asn^2, or Gln^2 (figure 6a). It is credited with assembling random sequence amide-bearing, polyanionic polypeptide chains, on a poly(A) template. [13-15] Replacement of the first tRNA adaptor implies competition took place with tRNA-IA^{Asn} $_{UUU}$ for AAA codons. Evolution of a type IA tRNA, specific for Asn^2, would distinguish it from the primordial adaptor, tRNA-ID$^{(Asp,Glu,Asn,Gln)}$ $_{UUU}$. Its replacement allowed polymerization of ordered sequence polypeptide chains directed by the NH_4^+ Fixers Code.

The large Asn-adaptor domain contains three sub-domains, rooted at Ser, Met, and Ala adaptors (figure 6a). Seven tRNA species branch in the Ser sub-domain, including three tRNA-I adaptors for Ser-derived aa, Cys^5, Gly^5, and Trp^{14}. There are five tRNA species in the Met-adaptor sub-domain and two in the Ala-adaptor sub-domain. tRNA-IB^{Phe} branched in region-3 of the LCA tree together with adaptors for Asp-derived aa (figure 4). This is in accord with descent from tRNA-IA^{Met} (figure 4, table 2). tRNA-IA^{Met} and tRNA-IB^{Phe} both, notably, read NUN column codons. The adaptor for Phe^{11} sibling, Tyr^{11}, likewise, branched in region-3. Adaptors for siblings, Ala^4 and Val^5, form the Ala sub-domain. The adaptor cluster pattern in the large tRNA tree domain (figure 6a) supports the proposed fragmentation of an enlarged pre-LCA Asp family into Asp, Pyr, Ser, and Shi families. [15]

Radiation of tRNA^{Asn}, leading to formation of the large tree domain, could reflect use of distinct tRNA-I subtypes for precursor and product aa in the Asp family (figure 6a). Aspartate^2 acylated a type ID tRNA^{Asp} and misacylated a type IA tRNA, for Asn, during product aa synthesis. This contrasts with the small domain, where Glu^2 acylated and misacylated a type ID tRNA. Use of different tRNA-I subtypes for precursor and product aa would aid molecular recognition, during tRNA^{Asn} radiation. In this connection, pre-LCA tRNA-dependent aa synthesis possibly reduced the number of (ribozymal) aminoacyl-tRNA-synthetases (r-aaRS) to two: those for the primordial aa precursors, Asp^2 and Glu^2. This would shift recognition away from the product aa itself to its tRNA adaptor/cofactor, whose base sequence evidently encoded aa identity and synthesis pathway, in the RNA World.

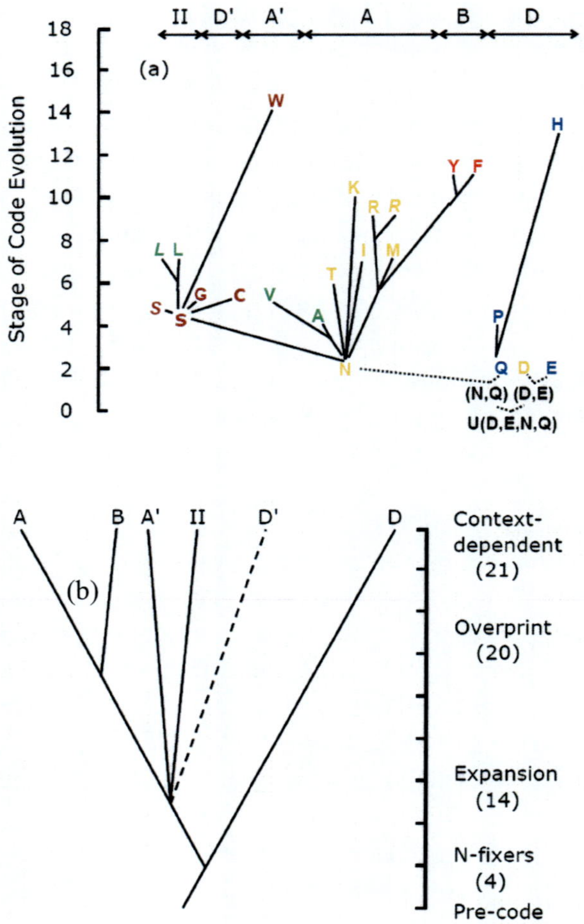

Figure 6. Pre-LCA tRNA diversification during code evolution. (a) Tree nodes join an aa adaptor with its putative ancestor, inferred from trace identity. The ordinal length of each branch corresponds to the stage of code evolution a tRNA species originated, estimated from the attached aa path-distance. A pre-code stage at the tree root involves a 'universal' type-ID tRNA adaptor for each NH_4^+ fixer aa. By stage-2, tRNA-ID adaptors specific for Asp^2, Glu^2, or Gln^2, and tRNA-IA adaptor for Asn, formed the NH_4^+ Fixers Code. tRNA adaptors for fourteen different aa, with type IA, IA', IB, ID' and II core structure, showed kinship with Asn-tRNA-IA. $Proline^4$ and His^{13} tRNA-ID descent from Gln-tRNA-D was favored by phylogenetic evidence. LCA tRNA partition into two deeply branching domains rooted at Asn-tRNA-IA, or the primordial NH_4^+ Fixer-tRNA-ID. Single letter abbreviations designate aa. Amino acid families are color-coded as in figure 4. ....., branch predates NH_4^+-Fixers Code (stage 2). ___, post-'NH_4^+-Fixers Code' branches. (b). Tree depicting evolution of tRNA types and subtypes. At stage-2, type IA tRNA (Asn^2-adaptor) is shown to diverge from tRNA-ID. tRNA with modified type IA core (Cys^5- and Val^5-adaptors) and type II tRNA (Ser^4- adaptors) diverged from an Asn-adaptor-like tRNA by stage-5. A dashed line indicates divergence from a tRNA-IA ancestor (presumptive ancestral Ser-tRNA) of the Gly-adaptors, with a type ID' core. Formation of type IB tRNA-I (aromatic aa adaptors) from tRNA-IA is placed at stage-11 of code evolution.

Coding Domains in the Standard Code

Figure 7 depicts the Standard Code as a patchwork of distinct domains. Each domain is a region of contiguous codons read by tRNA, possessing core structure homology, and charged by aa synthesized from a common precursor. They have a row-wise orientation within the code. This is anticipated from evidence for columnwise growth of the code (see Channeled diversification of tRNA species in the pre-LCA era).

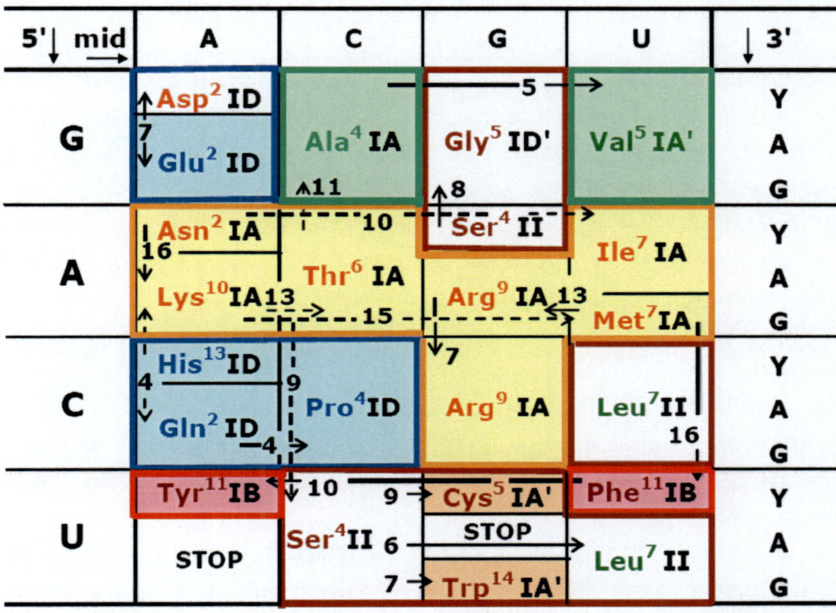

Figure 7. Location of domains, LCA tRNA tree regions and aa families in the Standard Code. Each domain contains contiguous codons read by tRNA adaptors, displaying core group homology, for same-family aa. The overlap between domains and tree regions indicate adaptors for same-family aa descended from a common ancestor. Roman numerals and upper-case letters denote tRNA type and subtype, respectively. An arrow connects an adaptor and its putative ancestor. Arrow number gives identity (quarts) between the conserved trace of adaptor and ancestor sequences. A dashed arrow connects adaptor to the ancestral Asn-adaptor, cognate with codons AAN. Adaptor/ancestor pairs occur mainly within code domains. Tan background marks Asp domain; blue, Glu domain; green, Pyr domain; yellow, Ser domain; and, red, Shi domain. Members of aa families are color-coded as in figure 4. Bold blue lines enclose tree region-1, with Glu family and Asp tRNA; green lines delineate region-2 for Pyr family adaptors; orange lines, region-3 for Asp-derived aa; brown lines, region-4 for Ser family and Leu; and red lines, region-5 for Shi family adaptors. This figure draws on results in figure 1 and 4, and table 2.

Codons for Asp family aa (Asp^2, Asn^2, Thr^6, Ile^7, Met^7, Arg^9, Lys^{10}) form the largest domain. It contains twenty triplets spread principally along the ANN row. Only fifteen are distinguishable during translation, however, because of 3'-pyrimidine (Y) degeneracy. [43] $Aspartate^2$ codons, GAY (figure 7), are not in the Asp family domain, as Asp^2 charges a type ID tRNA (figure 1), making it heterologous with the type IA adaptors of Asp-derived aa. CNN row codons, $CG^{A}{}_{G}$, assigned to Arg^9, are nearest-neighbor to ANN row Arg codons, $AG^{A}{}_{G}$. Hence, its CNN row codons ($CG^{A}{}_{G}$, CCY) are a contiguous extension of the Asp family domain. Aspartate family aa form on pathways that originate at oxaloacetate (OAA) in the citrate cycle, apart from Arg^9. [30, 31] Inclusion of Arg^9 in this family reflects a reaction between Asp^2 and Cit^7, at step 8 of synthesis, that makes Asp^2 the last direct precursor. [43]

Glutamate family aa (Glu^2, Gln^2, Pro^4, His^{13}) form on pathways originating at α-ketoglutarate (α-KG) in the citrate cycle, discounting His^{13} whose synthesis commences at ribose-5-phosphate in the pentose cycle. $Glutamine^2$ donates an amide group at step 9 of His^{13} synthesis (measured from the citrate cycle) [30], thereby recruiting this aa for the Glu^2 family. Five aa encoded by CNN row codons include three Glu family members (figure 7). All four Glu family aa have tRNA-ID adaptors, cognate with eight distinguishable, contiguous codons ($GA^{A}{}_{G}$, CAY, $CA^{A}{}_{G}$, CCY, $CC^{A}{}_{G}$).

Pyruvate-derived siblings, Ala^4 and Val^5, respectively charged related LCA tRNA-IA and –IA' cognate with GNN row codons, GCN, GUN (figure 7). A third Pyr-derived aa, Leu^7, acquired NUN column codons (UUR, CUN). $Leucine^7$ charges type II tRNA, placing Leu^7 codons outside the Pyr family code domain.

Serine family aa (Ser^4, Cys^5, Gly^5, Trp^{14}) derive from 3PG in the central trunk, with Trp^{14} the sole exception. A reaction between Ser^4 and indole, in the final step of synthesis, recruited Trp^{14} into the Ser family. [30] $Cysteine^5$ and Trp^{14} charged type IA related LCA tRNA cognate with UNN row codons, UGY and UGG, respectively (figure 1). A small Ser domain thus occurs in the code UNN row.

Aromatic aa, Phe^{11}, Tyr^{11}, form on the Shi pathway, originating with a reaction between phosphoenolpyruvate (central trunk) and erythrose-4-phosphate (pentose cycle). [30] They acylated type IB LCA tRNA cognate with codons UUY and UAY (figure 1, 2). A second small domain thus resulted in the code UUN row.

Thirty-two (32/45) distinguishable codons (STOP codons excluded) occur within the five domains of the Standard Code. Equivalently, forty-three (43/61) triplets allocated to aa occur within code domains. They encode fifteen siblings and one precursor (Glu^2) aa.

Domain codons are contiguous. Parameters determining the probability for chance occurrence of nearest-neighbor codons between each aa and other members within the same domain, synthesis family, or tRNA tree region appear in table 3. Codon contiguity was strongest within aa families. The combined probability [37] for codon contiguity among same-family aa equaled 5.09×10^{-8}, based on $\chi^2(df = 40)$ of 106.97. Codons read by tRNA, with designated aa identity, that branch within the same tree region (figure 4), or same code domain (figure 7), had a combined contiguity probability of 1.02×10^{-7} ($\chi^2(df = 40) = 104.9$), and 2.16×10^{-6} ($\chi^2(df = 32) = 82.9$), respectively. Closely contiguous codons thus encode aa of the same family, tRNA tree region, and code domain. Based on these probabilities, codons within each tree region were two-fold less contiguous than codons of aa families. While code domain codons were 42-fold less contiguous. Precursor heterogeneity (3PG, Pyr) among aa (Ser^4, Leu^7) charging type II tRNA and adaptor/ancestor heterogeneity within the Ser family ($tRNA\text{-}II^{Ser} \rightarrow tRNA\text{-}IA^{'Cys}$; $tRNA\text{-}II^{Ser} \rightarrow tRNA\text{-}IA^{'Trp}$; $tRNA\text{-}II^{Ser} \rightarrow tRNA\text{-}ID^{'Gly}$) lowered codon contiguity within code domains. Adaptor heterogeneity was attributed to replacement of an ancestral type I $tRNA^{Ser}$ by a type II $tRNA^{Ser}$, facilitating allocation of multiple codon sets to a single aa, during rapid expansion from the NH_4^+ Fixers Code (see Identity and source of LCA tRNA species).

Table 3. Codon contiguity in amino acid families, tRNA tree regions and code domains[1]

Amino acids2					Contiguity parameters[3]														
aa	Pre-cursor	Tree region	Code domain	Codons	a	b	x	n	-2lnp	a	b	x	n	-2lnp	a	b	x	n	-2lnp
					Precursor					Tree region					Code domain				
Glu	αKG	1	D	GAR	11	32	2	6	2.25	11	32	3	7	7.35	11	32	2	6	2.25
Gln	αKG	1	D	CAR	11	32	5	6	12.04	11	32	5	7	9.89	11	32	5	6	12.04
Pro	αKG	1	D	CCN	18	24	3	5	2.66	18	24	3	6	2.31	18	24	3	5	2.66
His	αKG	1	D	CAY	8	36	3	7	4.91	8	36	4	8	7.52	8	36	3	7	4.91
Ala	Pyr	2	A2	GCN	18	24	3	8	2.45	18	24	3	3	5.29	18	24	3	3	5.29
Val	Pyr	2	A2	GUN	18	24	8	8	15.80	18	24	3	3	5.29	18	24	3	3	5.29
Leu	Pyr	4	N	CUN/ UUR	22	18	3	6	2.23	19	21	3	9	3.30	-	-	-	-	0
Asp	OAA	1	N	GAY	8	36	3	10	2.46	8	36	3	8	4.26	-	-	-	-	0
Asn	OAA	3	A1	AAY	8	36	5	11	6.16	11	33	4	13	2.81	11	33	4	13	2.81
Thr	OAA	3	A1	ACN	18	24	6	9	6.34	18	24	8	11	7.56	18	24	8	11	7.76
Met	OAA	3	A1	AUG	8	36	5	11	6.16	8	36	5	13	6.84	8	36	5	13	6.84

Table 3. (Continued)

Amino acids[2]					Contiguity parameters[3]														
aa	Pre-cursor	Tree region	Code domain	Codons	a	b	x	n	-2lnp	a	b	x	n	-2lnp	a	b	x	n	-2lnp
					Precursor					Tree region					Code domain				
Ile	OAA	3	A1	AUY/A	13	30	6	10	4.70	13	30	5	12	3.53	13	30	5	12	3.53
Lys	OAA	3	A1	AAR	13	30	7	10	9.61	11	32	7	12	10.88	11	32	7	1	10.88
Arg	OAA	3	A1	AGR/ CGN	8	36	3	14	2.79	21	19	6	9	3.30	21	19	6	9	3.30
Cys	3PG	4	A3	UGY	7	37	4	8	8.68	7	37	4	13	4.96	7	37	1	1	3.68
Gly	3PG	4	N	GGN	17	25	3	6	2.42	17	25	3	11	3.52	-	-	-	-	0
Trp	3PG	4	A3	UGG	6	38	3	8	5.74	6	38	4	13	6.11	6	38	1	1	3.99
Ser	3PG	4	N	UCN/ AGY	21	20	3	5	2.17	21	20	5	10	2.54	-	-	-	-	0
Phe	Shi	5	B	UUY	8	36	1	1	3.41	8	36	1	1	3.41	8	36	1	1	3.41
Tyr	Shi	5	B	UAY	6	38	1	1	3.99	6	38	1	1	3.99	6	38	1	1	3.99

Table 3. (Continued)

Codon contiguity:	$\chi^2_{40} = -\Sigma\ 2 \ln p = 107.0$ $p = 5.09\text{x}10^{-8}$	$\chi^2_{40} = 104.9$ $p = 1.02\text{x}10^{-7}$	$\chi^2_{32} = 82.9$ $p = 2.16\text{x}10^{-6}$
Correlation[4]:	τ(precursor v. region) = 0.92, $p = 7.0\text{x}10^{-9}$ τ(region v. domain) = 0.72, $p = 8.3\text{x}10^{-6}$ τ(precursor v. domain) = 0.68, $p = 1.5\text{x}10^{-5}$		

[1] Codons differing by a 3'-pyrimidine were treated as indistinguishable. STOP codons were omitted.

[2] Tree regions are from figure 4. Code domains are from figure 7. N, no domain contained these amino acid adaptors. A1, A2 and A3 refer to three distinct domains with tRNA adaptors in the type IA group (includes IA').

[3] Nearest-neighbor codon frequencies were assessed by hypergeometric probabilities, $p(0 < x \leq n)$, and $p(x = 0) \equiv 1$, for each aa versus members of same family, each tRNA within same tree region and for each aa within a code domain. a, number of triplets a single base substitution from codons of a specified aa. b, number of codons more than one substitution from codons for this aa. x, number of codons read by adaptors for related aa one substitution from a specific aa codons. n, total number of codons assigned to the designated aa. Codon contiguity probabilities for each aa were combined by Fisher's method. [37]

[4] Kendall correlation adjusted for tied ranks. Probabilities were from the normal distribution. [42]

Amino acid families, code domains, and tRNA tree regions overlap extensively in the Standard Code (figure 7). Amino acid precursors correlated with aa identity of adaptors in the same tRNA tree region (figures 4, 5) to produce a Kendall rank coefficient [42], τ, of 0.92, p = 7.0×10^{-9} (table 3). Same-family and same-domain aa corresponded with a τ coefficient of 0.68, p = 1.5×10^{-5}. Amino acid identity of same-tree region tRNA and same-domain aa had a τ coefficient of 0.72, p = 8.3×10^{-6}.

Code structure has provided compelling evidence that tRNA diversification from a primordial adaptor, growth of aa synthesis pathways outward from central metabolism, and codon assignments during code evolution were coordinated. Amino acids synthesized from a common precursor generally charged tRNA descended from a common ancestor and cognate with codons within same code row. A dual tRNA role, as cofactors in aa synthesis and adaptors in translation, is credited with coordinating aa pathway growth and code evolution with pre-LCA tRNA diversification. Columnwise expansion of the code from NAN → NCN → NGN → NUN [13, 15] is attributed with producing the row-wise orientation of code domains (figure 7). 5'-Base invariance among codons of same-family aa, as noted, also conforms with this growth pattern. [17] Growth of the code column-by-column reduced the risk of a potentially lethal [44] mutation to an unassigned triplet, within a template sequence; lowering it initially to the chance of a codon mid-base substitution. [13, 15] Code structure also implied code expansion was subject to a hydrophobic attractor [15] and error-minimization, targeting codon mid-base misreading. [18] Overprinting in code rows with short-path sibling aa (Lys^{10}/Asn^{2}; His^{13}/Gln^{2}; Trp^{14}/Cys^{5}), by post-expansion additions to the code (Arg^{9}, Lys^{10}, Phe^{11}, Tyr^{11}, His^{13}, Trp^{14}), indicated the dual cofactor/adaptor role of pre-LCA tRNA channeled codon assignments throughout code evolution. [13, 15]

Free tRNA Attachment Site in Extant Amino Acid Intermediates

Retention of a free tRNA attachment site by intermediates in extant aa synthesis pathways is seen here as evidence supporting participation of tRNA cofactors in pre-divergence aa synthesis. All aa intermediates are noted to bear a free α–carboxyl group, except those in the His^{13} pathway and at step-12 and -13 in Trp^{14} synthesis. [30, 31]

The carboxyl vicinal to a keto group in OAA, α-KG, and Pyr is the source of the α-carboxyl in thirteen aa produced by the citrate cycle: Asp^{2}, Asn^{2}, Thr^{6}, Met^{7}, Ile^{7}, Arg^{9}, Lys^{10}, Glu^{2}, Gln^{2}, Pro^{4}, Ala^{4}, Val^{5}, Leu^{7}. Five aa originate from the central trunk and they contain an α-carboxyl from 3PG, or phosphoenolpyruvate: Ser^{4}, Cys^{5}, Gly^{5}, Phe^{11}, Tyr^{11}. $Tryptophan^{14}$ intermediates bear a phosphoenolpyruvate α-carboxyl until step- 12, when it is removed on decarboxylation of o-carboxyphenyl-amino-1-deoxyribulose-5-phosphate. [30] Two steps later, indole reacts with Ser^{4}. $Tryptophan^{14}$ thereby acquired the α-carboxyl of Ser^{4} precursor, 3PG. $Histidine^{13}$ intermediates lack an α-carboxyl until oxidation of histidinol, in the final step of synthesis.

α-Amine (α-imino) addition, by contrast, generally takes place in the final steps of an aa synthesis pathway. In the synthesis of seventeen aa, α-amine addition occurs at the next-to-penultimate step, or later: Asp^{2}, Asn^{2}, Ile^{7}, Lys^{10}, Glu^{2}, Gln^{2}, Pro^{4}, His^{13}, Ser^{4}, Cys^{5}, Gly^{5}, Trp^{14}, Ala^{4}, Val^{5}, Leu^{7}, Phe^{11}, Tyr^{11}. A departure from this pattern arises in the synthesis of siblings Thr^6 and Met^7: all their intermediates possess a free α amine. Since Met^7, or f-Met, initiate extant translation, the α-aa intermediates in the Met pathway conceivably performed this function, before Met^7 had appeared.

An α-amine contributed by an aa precursor is lost in the synthesis of Pro^4, Ile^7 and Lys^{10}, only to be re-acquired in the final steps of synthesis. During Arg^9 synthesis, deacetylation of N-acetyl-ornithine, at step-6, unmasks an α-amine contributed by its primary precursor, Glu^2. [31]

α-Amine addition thus produces a scattered path-length distribution, in contrast to the α-carboxyl distribution (figure 8). The timing of α-amine and α-carboxyl addition differs significantly; $p = 1.45 \times 10^{-8}$, according to a Kolmogorov-Smirnov goodness of fit test. [42]

The ubiquity of a free α-carboxyl among intermediates, in extant aa synthesis pathways, conforms with this group being masked by attachment to a tRNA cofactor, prior to the protein takeover of aa synthesis pathways, in the pre-LCA era. Late addition of the α-amine group would then provide an error-avoidance strategy, preventing translation of an intermediate attached to a cofactor/adaptor tRNA.

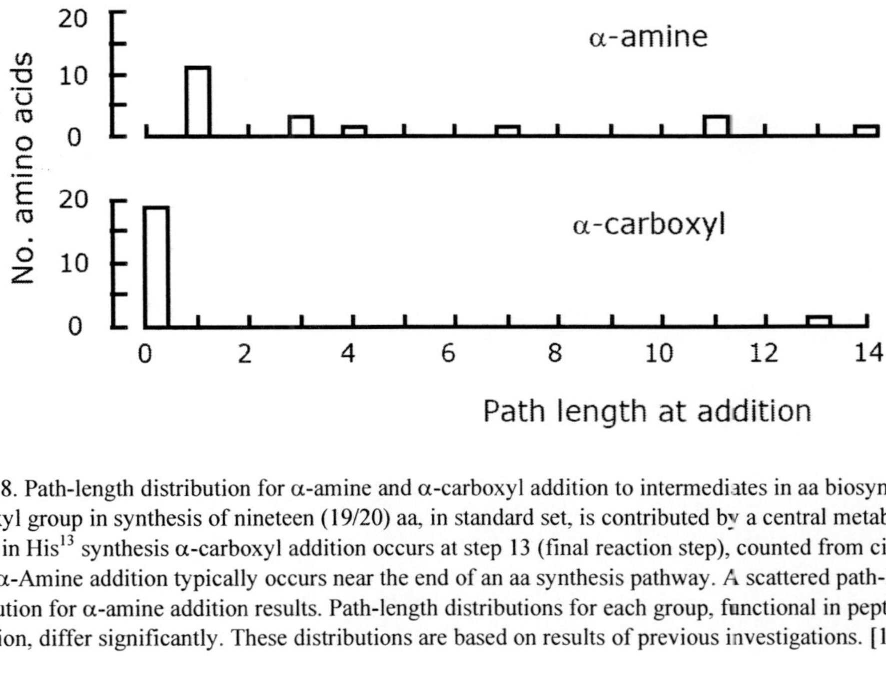

Figure 8. Path-length distribution for α-amine and α-carboxyl addition to intermediates in aa biosynthesis. α-Carboxyl group in synthesis of nineteen (19/20) aa, in standard set, is contributed by a central metabolite at step 0; in His^{13} synthesis α-carboxyl addition occurs at step 13 (final reaction step), counted from citrate cycle. α-Amine addition typically occurs near the end of an aa synthesis pathway. A scattered path-length distribution for α-amine addition results. Path-length distributions for each group, functional in peptide bond formation, differ significantly. These distributions are based on results of previous investigations. [13, 15]

Chemical Homology within Amino Acid Families

Evidence presented in this section relates to the possibility that significantly elevated chemical homology among same-family aa accounted for them charging homologous tRNA, cognate with triplets in the same code domain.

Table 4 lists thirty-seven Sneath coefficients [47] for within-family aa chemical homology and one hundred and fifty-three coefficients characterizing between-family homology. Within-family pairs achieve a mean Sneath coefficient of 0.405 ± 0.032 (8 ±

SEM). This exceeds the between-family mean coefficient of 0.368 ± 0.012 per aa pair. The excess is not significant, however; t(df = 1) = 2.06, p = 0.144, in a one-tail t test. [36]

Occurrence of a non-significant elevation in chemical homology among same-family aa discounts homology as the source of code structural regularities, which were demonstrated to be highly correlated with aa metabolic origin (table 3).

Table 4. Amino acid homology within and between families[1]

Family		Glu			Pyr			Asp							Ser				Shi	
	aa	Gln	Pro	His	Ala	Val	Leu	Asp	Asn	Thr	Met	Ile	Arg	Lys	Ser	Cys	Gly	Trp	Phe	Tyr
Glu	Glu	.69	.00	.20	.24	.33	.22	.84	.58	.22	.40	.37	.38	.44	.31	.22	.05	.09	.22	.26
	Gln		.17	.41	.36	.42	.50	.49	.59	.34	.70	.45	.56	.55	.50	.48	.16	.35	.46	.37
	Pro			.17	.55	.47	.43	.00	.17	.33	.27	.42	.16	.30	.32	.32	.50	.21	.21	.18
	His				.32	.30	.45	.20	.46	.21	.35	.37	.40	.42	.34	.29	.18	.48	.61	.50
Pyr	Ala					.68	.64	.16	.28	.42	.42	.59	.32	.43	.48	.58	.66	.22	.36	.21
	Val						.79	.28	.40	.55	.47	.84	.31	.42	.44	.43	.44	.20	.38	.20
	Leu							.43	.51	.43	.52	.89	.36	.49	.41	.40	.38	.37	.57	.35
Asp	Asp								.64	.25	.29	.34	.24	.25	.33	.24	.02	.03	.17	.29
	Asn									.49	.52	.46	.43	.40	.58	.49	.19	.32	.34	.39
	Thr										.41	.49	.26	.22	.67	.49	.40	.18	.25	.29
	Met											.49	.52	.48	.48	.61	.15	.36	.47	.31
	Ile												.34	.48	.36	.35	.37	.29	.49	.27
	Arg													.73	.32	.32	.15	.29	.34	.41
	Lys														.29	.27	.31	.30	.38	.29
Ser	Ser															.61	.32	.24	.38	.35
	Cys																.29	.19	.29	.22
	Gly																	.14	.26	.16
	Trp																		.74	.57
Shi	Phe																			.73
	Tyr																			

[1] Chemical homology coefficients for aa.[47] Coefficients for same-family aa appear in triangles. Coefficients outside the triangles specify homology between aa from different families. Among thirty-seven within-family aa pairs, the mean homology coefficient was 0.405 ± 0.032 (8 ± SEM). The between-family mean was 0.368 ± 0.012 for 153 pairs. A one-tail t-test established the difference in means was not significant; t(df = 1) = 2.057, $p = 0.144$.

Encoding by Aminoacyl-tRNA Synthetases

Two distinct classes of enzymes (p-aaRS), from the pre-LCA era, specifically attach an activated aa to a tRNA molecule in extant systems. [46] We may enquire: (1) At what stage in pre-LCA evolution did each class of p-aaRS begin matching aa and base triplets? And, (2) how were aa and triplets matched before the first p-aaRS appeared?

Class I and II p-aaRS are themselves the product of code-directed translation, so question (1) can be recast as follows: at what stage after the origin of the genetic code did each p-aaRS class appear and did this contribute to shaping the code? Residue 'code age' (superscripts) in Rossman fold signature segments $H^{13}I^{7}G^{5}H^{13}$ and $K^{10}M^{7}S^{4}K^{10}$ of class I aaRS [47] indicate class I synthetases originated no earlier than the stage-13 code. [13-15] Motifs P^{4}, $F^{11}R^{9}$-E^{2}, and G^{5}-G^{5}-G^{5}-$E^{2}R^{9}$ in class II aaRS apparently exclude formation of class II synthetases prior to the stage-11 code. The requirement for an advanced code at the origin of p-aaRS can be inferred also from frequency-shift estimates [22] for the time-of-entry into the genetic code of synthetase residues. Presently, there is no evidence that p-aaRS retain function when reconstructed with a reduced aa alphabet, as demonstrated in some proteins. [41, 48]

The distribution of p-aaRS among members of aa families agrees with the late origin scenario. Table 5 shows that aaRS-I and -II are randomly distributed among aa families. Since code structure and aa synthesis are closely linked (table 3), the absence of a correlation between aa metabolic source and p-aaRS class effectively excludes the possibility that synthetases shaped the genetic code. Code structure, on the orher hand, evidently did affect the distribution of p-aaRS classes over the code.

Table 5. Distribution of class I and II aminoacyl-tRNA synthetases in amino acid families[1]

Amino acid	Synthetase I	Synthetase II	p^2	Amino acid	Synthetase I	Synthetase II	p
	Aspartate family				*Pyruvate family*		
Asp^{2}		+		Ala^{4}		+	
Asn^{2}		+		Val^{5}	+		
Thr^{6}		+		Leu^{7}	+		
Met^{7}	+			Distn.	2	1	0.375
Ile^{7}	+						
Arg^{9}	+						
Lys^{10}		+					
Distn.	3	4	0.273				
	Glutamate family				*Serine family*		
Glu^{2}	+			Ser^{4}		+	
Gln^{2}	+			Cys^{5}	+		
Pro^{4}		+		Gly^{5}		+	
His^{13}		+		Trp^{14}	+		
Distn.	2	2	0.375	Distn.	2	2	0.375
					Shikimate family		
				Phe^{11}		+	
				Tyr^{11}	+		0.50

Global distribution: synthetase classes, $\chi^2_{10} = 9.86$, $p = 0.452$

[1] Synthetase aa specificity is from Eriani et al. [47]

[2] *p*, probability of principal synthetase in each aa family. Global probability combined family probabilities by Fisher's method. [37]

Physicochemical differences between aa substrates of p-aaRS-I and -II have been examined. [49] Class I aaRS aa (Glu^{2}, Gln^{2}, Cys^{5}, Val^{5}, Met^{7}, Ile^{7}, Leu^{7}, Tyr^{11}, Trp^{14}) have a significantly higher mean mol. wt. than those of aaRS-II (Asp^2, Asn^2, Ala^4, Pro^4, Ser^4, Gly^5, Thr^6, Lys^{10}, Phe^{11}, His^{13}); 150.1 ± 8.9 ($8 \pm$ SEM) versus 123.3 ± 9.0, $p = 2.4 \times 10^{-2}$ ($t(df = 1) = 26.1$) in a one-tail t-test. [36] Both sets of aa overlap substantially, however: aaRS-I aa range in mol. wt. from 117 (Val^5) to 204 (Trp^{14}), and aaRS-II aa range from 75 (Gly^5) to 165 (Phe^{11}). Seven aaRS-I paralogs (7/10) act on aa whose mol. wt. is actually smaller than the aaRS-II upper limit, 165. Conversely, six aaRS-II paralogs (6/10) act on aa whose mol. wt. exceeds the aaRS-I lower limit, 117. Each set of p-aaRS paralogs, therefore, acts on a range of aa, whose size virtually spans the complete spectrum of aa molecular weights.

An initial difference between aa substrates of the first p-aaRS, of each class, plausibly led to the difference in mean mol. wt. among aa substrates of each set of p-aaRS paralogs (table 5). This can be simply visualized as p-aaRS evolution radiating from a primordial synthetase, which acted on an aa of nearly average mol. wt. Candidate substrates for the first aaRS-I are: Glu^2, Gln^2, and Met^7, whose mol. wts. (146-149) cluster close to the class I average (150). Judging from the ubiquity of tRNA-dependent Gln^2 synthesis in prokaryotes, direct acylation of $tRNA^{Gln}$ originated late. [52] This discounts GlnRS as the first p-aaRS-I. Furthermore, phylogenetic analysis indicates GlnRS descended from GluRS. [46] Also, Met-RS is a protein dimer and, seemingly, too complex for the primordial aaRS-I. By contrast, GluRS-I is a monomer of the standard p-aaRS-I form and taxonomic evidence shows it is evolutionarily deep. [46] Thus, GluRS can be tentatively identified as the first p-aaRS-I.

$Aspartate^2$, Asn^2, and Thr^6 have mol. wts. (119-132) that span the mean aa mol. wt. for p-aaRS-II substrates (123). Their synthetases are homodimers. [46] Use of a tRNA cofactor is again a feature of the amide aa synthesis path, indicative of a late (post-divergence) transition to direct tRNA acylation. [50] AspRS is evolutionarily deep and ancestral to the amide aa synthetase. It is also closer to the canonical p-aaRS-II structure than ThrRS. [48] It appears likely, therefore, that AspRS was the first aaRS-II.

tRNA-dependent amide-aa synthesis pathways persisted into the post-divergence era and still occur in some prokaryote species. These pathways plausibly represent the protected root (figure 6) of a once extensive network of pre-LCA tRNA-dependent aa synthesis pathways. An alternative interpretation suggesting the tRNA-dependent pathways of amide-aa indicates a late entry into the code [50] is at variance with the path-distance model of code evolution, since Asn and Gln form on 2-step paths. [13-15] Both precursor aa, Asp and Glu, form on a one-step path and, unlike their derived amide aa, each path did not apparently involve a tRNA cofactor. Initially, a primordial ribozymal diacid synthesase (r-aaRS) evidently charged both $tRNA^{dicarboxyl\text{-}aa}$ and $tRNA^{amide\text{-}aa}$. Reliance on an r-aaRS plausibly made $tRNA\text{-}ID^{Asp}$ and $tRNA\text{-}ID^{Glu}$ acylation a target for takeover by the primordial class II and I synthetase, respectively. The difference in mol. wt. between homologues Asp (132) and Glu (146) is consonant with it producing the disparity in mean mol. wt. among aa substrates of aaRS-II (123.3 ± 9.0) and -I (150.1 ± 8.9).

All aa, with NCN column codons (Ala^4, Thr^6, Pro^4, Ser^4), acylate the 3'OH of tRNA A76, in the presence of an p-aaRS-II. [51] On the other hand, amino acid encoded by NUN triplets (Val^5, Ile^7, Met^7, Leu^7, Phe^{11}) esterify the 2'OH of A76, with a p-aaRS-I, or anomalous class II aaRS for Phe, catalyzing the reaction. These NCN and NUN clusters in p-aaRS distribution closely approach statistical significance, $p = 5.8 \times 10^{-2}$ for combined column-probabilities [37], with $\chi^2(df = 8) = 15.1$. NUN triplets encode more hydrophobic aa than

NCN triplets; change in residue mean free energy on transfer from an aqueous to non-polar (dielectric constant, 2.0) solvent being, respectively, -3.0 and -0.8 kcal/mol. [13] Overall aa substrates of p-aaRS-I paralogs have, however, only slightly lower residue transfer free energy than aaRS-II paralogs: 0.95 ± 1.73 (8 ± SEM) versus 1.79 ± 1.42 kcal/mol, respectively; a non-significant difference in a one-tail t test [36], p = 0.36 (t(df = 1) = 1.60). Residue hydrophobicity among each set of p-aaRS paralogs overlap extensively: class I range, 12.3 (Arg^9) to -3.4 (Met^9) kcal/mol, and class II range, 9.2 (Asp^2) to -3.7 (Phe^{11}) kcal/mol. [13]

Synthetase evolution appears, from present considerations, to have commenced with appearance of GluRS-I and AspRS-II. They putatively replaced ribozymal synthetases for these diacid aa precursors of other aa, deep within the pre-LCA era, when the synthesis of product aa occurred homotopically on tRNA cofactors. [2] Thereafter, phylogenetic evidence [48] implies a p-aaRS-I, ancestral to synthetases for Val^5, Ile^7, and Leu^7 (mol. wt., 117-131), initially loaded a branched-chain, highly hydrophobic aa, onto a tRNA, cognate with a set of NUN column triplets. Subsequently, it diversified to form synthetases specific each aa homologues encoded by NUN triplets. In a parallel development, a p-aaRS-II, ancestral to synthetases for Ser^4, Thr^6, and Pro^4 (mol. wt., 105-119), initially loaded a comparatively small, slightly hydrophobic aa onto a tRNA, cognate with some NCN column triplets. This synthetase, likewise, diversified to form p-aaRS for aa homologues encoded by NCN triplets. Column-clusters of p-aaRS, [49] from the perspective of the path-distance model of code evolution, represent the imprint of code structure [13, 15] on synthetase evolution, late in the pre-LCA era. Synthetase-I and -II clusters centered on NUN and NCN encoded aa, respectively, are congruent with the size-difference of aa substrates (Glu, Asp) for the proposed primordial synthetases, GluRS-I and AspRS-II. Central to this interpretation is the significance of the initial state in the radiation of pre-LCA enzymes.

In brief: (1) p-aaRS-I and II paralogs each display specificity for nearly the full spectrum of residue sizes and hydrophobicities. (2) They have an equal number (ten) of aa substrates (table 5). (3) p-aaRS class, and subclass, specificity for aa is phylogenetically invariant, with the notable exception of euryarchaea LysRS-I. [46] (4) aaRS-I and -II universally conserve residues with an advanced 'code age'. From (3) and (4), p-aaRS constitute a set of pre-LCA era proteins, which originated when the genetic code was apparently complete, or nearing completion. Based on (1), p-aaRS-I and -II paralogs were virtually unconstrained in the range of substrate aa. And, (2) the 10/10 split of aa between class I and II synthetases represents the most probable outcome, with aaRS-I and -II evolving comparable rates during the evolution of direct tRNA acylation in the pre-LCA era.

CONCLUSION

Five domains of contiguous codons encoded synthetically related aa that charged tRNA, with core structure homology, in the genetic code of the LCA. tRNA that diversified from a common ancestor are credited with coordinating the growth of aa synthesis pathways, originating from a common precursor in central metabolism, with assignment of base triplets, from the same code row, to successive generations of incoming aa. This strongly indicates pre-LCA tRNA adaptor molecules were utilized as cofactors in early aa synthesis. The acquisition of phylogenetically related type II tRNA by Ser^4 and Leu^7, whose isoacceptors were cognate with multiple sets of codons, point to both aa having a common precursor in the pre-LCA era. Elevated base identity of tRNA-II^{Ser} with tRNA-I^{Asn} suggested the ancestral Ser^4 adaptor was a type I tRNA. Its replacement then accounts for adaptor heterogeneity between Ser^4 and the Ser-derived aa Cys^5, Gly^5 and Trp^{14}.

Pre-LCA tRNA separated into two deeply branching phylogenetic clusters. tRNA for fifteen aa branched in the larger cluster. They displayed elevated identity with the type-IA tRNA Asn-adaptor. The aa identity of these tRNA conformed with misacylation of Asp^2 onto tRNA descended from an Asn-tRNA, during pre-LCA synthesis of Asp-derived aa. Occurrence of Pyr, Ser and Shi subclusters support existence of an enlarged pre-LCA Asp family, including the Pyr, Ser and Shi families, prior to its fragmentation on pathway rearrangement following displacement of tRNA cofactors by enzyme-catalyzed synthesis. The smaller, earlier cluster contained type-ID tRNA adaptors for both diacid aa, Gln^2 and Glu family siblings Pro^4 and His^{13}. A pre-informational tRNA-ID, cognate with triplet AAA and charged by NH_4^+ fixer aa, was placed at the root of the tRNA tree. Ribozymal synthetases r-AspRS and r-GluRS evidently charged pre-LCA tRNA with precursor aa (Asp^2, Glu^2) in pre-LCA tRNA-dependent aa synthesis. The evolutionary depth of GluRS and AspRS conform with them being the first class-I and -II synthetase enzymes, respectively. Column-clusters of p-aaRS-I and -II, within the code, then reflect the imprint of code structure on synthetase evolution.

METHODS

tRNA Sequences

Sequences of forty-eight tRNA species were obtained in a search of the Bayreuth database (www.uni-bayreuth.de/departments/biochemie/tRNA/). [28] 1100 base sequences, around 73 bases each, from sources in Archaea, Bacteria, and Eukarya were examined in this survey (Ref. 52 gives tRNA sources and access numbers). A consensus sequence was determined for tRNA, with designated aa identity and codon specificity, for sources from each species domain. The trace of an ancestral tRNA, predating species divergence from the LCA, was reconstructed from sites with an identical base in the consensus sequence representative of a tRNA species in each of the three domains of life. Residues conserved across species domains in consensus sequences for low potential ferredoxin and proteolipid helix-1 were previously reported to match corresponding residues in the LCA monomer sequence, inferred from phylogenetic trees, with a 93-100 per cent accuracy. [14] The computational efficiency achieved by using consensus sequences to filter-out post-divergence variability incurred a cost: the number of sites resolved in the LCA molecule was reduced. A half to three-quarters of the sites obtained by phylogenetic analysis was resolved. More complete reconstruction of LCA sequences with phylogenetic methods raises the possibility that a future investigation may extend the results presented in this chapter.

Sites conserved by virtually all tRNA species in this survey were credited to a primordial tRNA adaptor that predated tRNA diversification, deep within the pre-LCA era. The remaining, non-universal sites, strictly conserved across species domains in a given tRNA species, provided information relating to pre-LCA tRNA diversification (figure 9).

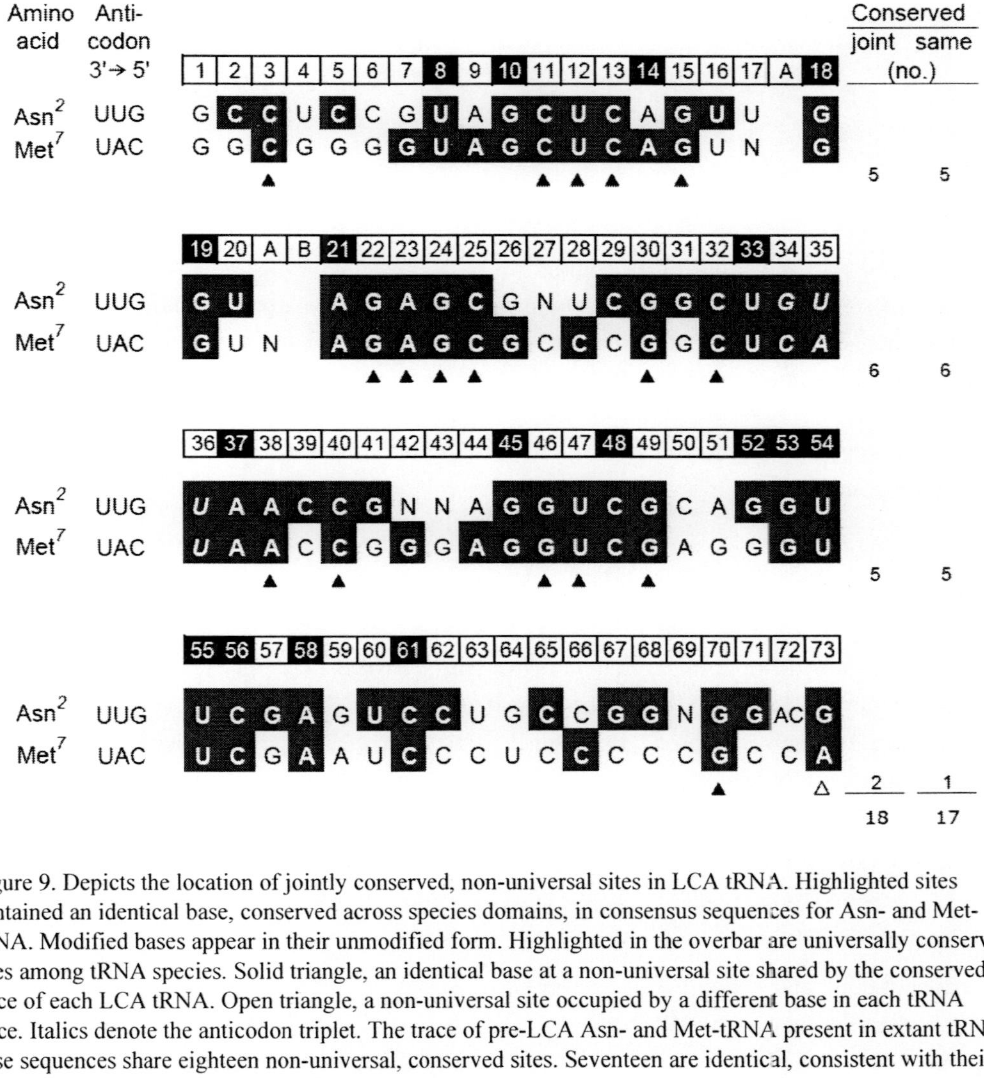

Figure 9. Depicts the location of jointly conserved, non-universal sites in LCA tRNA. Highlighted sites contained an identical base, conserved across species domains, in consensus sequences for Asn- and Met-tRNA. Modified bases appear in their unmodified form. Highlighted in the overbar are universally conserved sites among tRNA species. Solid triangle, an identical base at a non-universal site shared by the conserved trace of each LCA tRNA. Open triangle, a non-universal site occupied by a different base in each tRNA trace. Italics denote the anticodon triplet. The trace of pre-LCA Asn- and Met-tRNA present in extant tRNA base sequences share eighteen non-universal, conserved sites. Seventeen are identical, consistent with their close pre-LCA kinship. tRNA sequences are from figure 1.

LCA tRNA Types and Subtypes

Types and subtypes of LCA tRNA were identified from bases in the D stem (sites 8-15, 20A-26), its surround, and variable loop (sites 41-48) of sequences in figure 1. Bases at these sites form a set of interactions that had been shown to characterize the core structure of each kind of tRNA L form in thermophilic archaea and related species. [29] Generic sequences for each type, and subtype, found in LCA tRNA were constructed from consensus sequences for all tRNA attributed with the same core structure, sites not strictly conserved across the three species domains in a tRNA species were determined by majority rule.

Evidence that a family of synthetically related aa preferentially charged one type/subtype of LCA tRNA was obtained from core homology among adaptors for same-family aa. The probability, p, for chance occurrence of the frequency exhibited by the principal type/subtype of tRNA adaptor in an aa family was obtained from the binomial distribution,

$$p(x) = \binom{n}{x} \nu^{x}(1 - \nu)^{n-x} \tag{1}$$

x is the number of aa with an adaptor having the principal form of core structure, in a family with n members; each kind of aa charged only one type/sub-type of tRNA. ν signifies the normalized frequency of the principal type/subtype of tRNA, among adaptors for the twenty standard aa. The term in brackets, in Eqn. (1), is a binomial coefficient. Probabilities for each family were combined by method of Fisher, [37] to obtain the global probability for tRNA homology.

An analogous procedure was adopted to determine the degree of homology among p-aaRS within aa families.

Phylogenetic Analysis

Jukes-Cantor distances.- Phylogenetic kinship between tRNA paralogs in the LCA was assessed from Jukes-Cantor distances [40] determined at jointly conserved, non-universal sites (figure 9). The distance, d_{ij}, between the trace of two LCA tRNA species, conserved in extant sequences (figure 1),

$$d_{ij} = \frac{-3}{4} \ln\left(1 - \frac{4\, s_{ij}}{3\, n_{ij}}\right) \tag{2}$$

where anticodon and variable loop sites are excluded. s_{ij} is the number of sites with dissimilar bases and n_{ij} the total number of relevant sites. i and j refer to tRNA species. The logarithmic argument was non-positive in 5 per cent (60/1176) of sequence comparisons. This reflects a chance excess in the difference ratio, $s_{ij}/n_{ij} > 0.75$, between randomly related sequences. In this event, the argument was reset to 1×10^{-3}, yielding a difference ratio, $s_{ij}/n_{ij} \approx 0.749$.

Adaptation of the mutation distance metric to quantify differences between paralogs at mutually conserved sites provides a means to apply phylogenetic calculations to pre-LCA era molecular evolution. The distance obtained represents an estimate of the actual distance between both molecular species when they existed in the LCA, since it is based on the

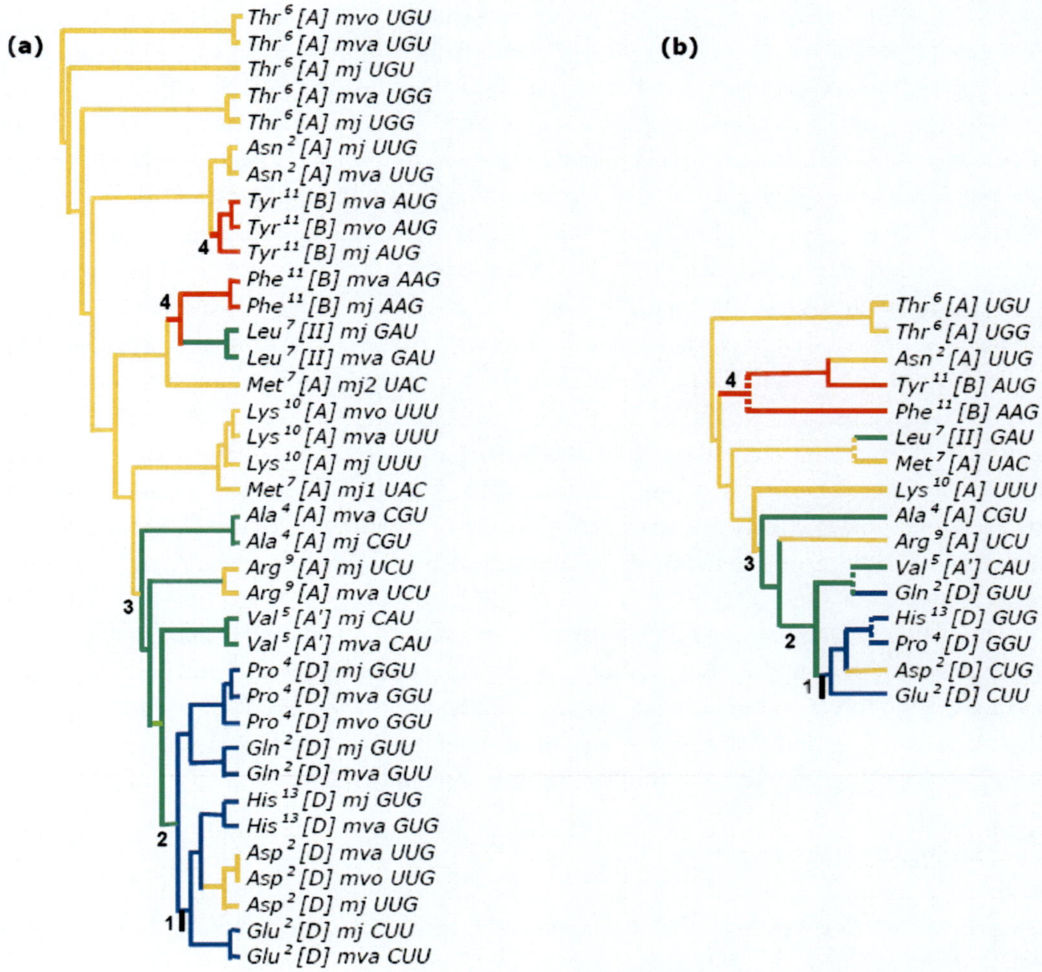

Figure 10. Compares cladograms obtained from analysis of the full-sequence and conserved-trace in *Methanococcus* tRNA. a. Sequence cladogram for sixteen tRNA species charged by fifteen kinds of aa. b. Conserved-trace cladogram for non-universal sites in tRNA conserved between *Methanococcus* species. Phylogenetic relations are shown to be highly conserved under consolidation of tRNA sequences from different *Methanococcus* species and restriction to jointly-conserved, non-universal tRNA sites. Amino acid identity stacking order in both cladograms had a Kendall correlation coefficient, [42] $\tau = 0.97$ ($p = 8.8 \times 10^{-8}$). Broken lines mark nodes that differ from the standard, full-sequence cladogram. Phylogenetic analysis was performed by neighbor-joining. [39] tRNA were from *M. jannaschii* (mj), *M. vanielii* (mva) and *M. voltae* (mvo). tRNA core structure group and anticodon bases (3' → 5') are specified. Branch color indicates family of acylating aa, as indicated under figure 4. *Methanococcus* tRNA sequences were from the Bayreuth database.

differences over the available sample of sites, rather than the complete set. Comparing cladograms based on the full-sequence versus a mutually-conserved-trace distance in tRNA from archaeal species illustrates the efficacy of this procedure. Trace distances are seen to conserve the stacking order of aa identity in the Methanococcus tRNA tree in figure 10. The Kendall rank correlation for the stacking order in both trees was, [42] $\tau = 0.97$, $p = 8.8 \times 10^{-8}$. Consolidation of tRNA ortholog sequences, to obtain the conserved trace, pruned the number of branch tips by approximately half (16/37). It is standard practice to align post-LCA sequences whose lengths differ, as a result of insertions and deletions. Subsequent phylogenetic analysis is, however, routinely based on the complete set of sites in the standard sequence and this distinguishes it from the present approach with pre-LCA sequences.

Neighbor-joining calculations, [39] involving trace Jukes-Cantor distances (Eqn. 2) for LCA tRNA (figure 1), led to a tree depicting pre-LCA tRNA diversification. The tRNA tree root was located using attached aa path-distance to estimate the time-order of events in pre-LCA tRNA diversification; [13, 15] analysis of tRNA paralogs from a single source (LCA) precluded use of outgroups. The distribution of aa identity, core structure type, and anticodon bases within the tRNA tree obtained was evaluated statistically to identify clusters.

Identity.- Jukes-Cantor distance solely reflects the fraction of dissimilar sites between two sequences, s_{ij}/n_{ij} (Eqn. 2). As this measure has no direct dependence on sequence size, it provides an intensive measure of sequence relatedness. When size differences, n_{ij}, are negligible, it is an intuitively reasonable indicator of evolutionary relatedness; distance being an inverse expression of relatedness. The trace of LCA tRNA conserved in extant sequences, however, may differ widely between tRNA species. In the present survey, the number of sites, n_{ij}, mutually shared between in the conserved trace of a pair of LCA tRNA species varied from 0 to 36 sites.

The matching probability (Eqn. 1) depends on both the number of identical sites, x, and total number of mutually conserved sites, n_{ij}, in ancestral sequences. Unlike Jukes-Cantor distances, therefore, it provides an extensive measure of sequence relatedness. This is relevant to investigations on pre-LCA evolution reliant on the conserved trace of ancestral molecules. The logarithm (base 4) of the matching probability (Eqn. 1),

$$I_{ij} = -\log_4 p(x_{ij}) \qquad (3)$$

furnishes an additive measure of sequence relatedness. n_{ij}, the total number of non-universal sites, conserved jointly by tRNA species i and j (Eqn. 2) corresponds to n in Eqn. 1. The number of identical sites is, $x = x_{ij} = n_{ij} - s_{ij}$. There is a random chance, ν, that a given kind of base will occur at any site equal to ¼ in a ribopolynucleotide sequence (A, G, C, U).

Sequence identity, I_{ij}, is expressed in quaternary units (quarts): $I_{ij} = -\log_4 p(x_{ij} = n) = -\log_4 4^{-n} = n$ quarts, when identical bases occur at n relevant sites. Quaternary identity has the form of a measure of uncertainty (defined within a multiplicative constant), with well known mathematical properties. [53] A conserved tRNA trace of n sites with an identity of n quarts implies that an uncertainty (entropy) corresponding to n choices among four kinds of bases is removed, on matching with a reference trace that specifies the identity of all sites in the test trace.

It may be noted that a 4^n-fold difference exists in matching probability among tRNA paralogs with n and 2n identical sites (s = 0): $p(n) = 4^{-n}$ and $p(2n) = 4^{-2n}$, giving $p(n)/p(2n) = 4^{-n}/4^{-2n} = 4^n$. A large sequence-size dependent difference in tRNA relatedness thus occurs. In contrast, Jukes-Cantor distances are equal for both pairs of tRNA. From Eqn. 2, $d(n, s = 0) = 0$, and $d(2n, s = 0) = 0$. Jukes-Cantor distance, according to this, overvalues sequence relatedness between the species with a smaller jointly conserved trace.

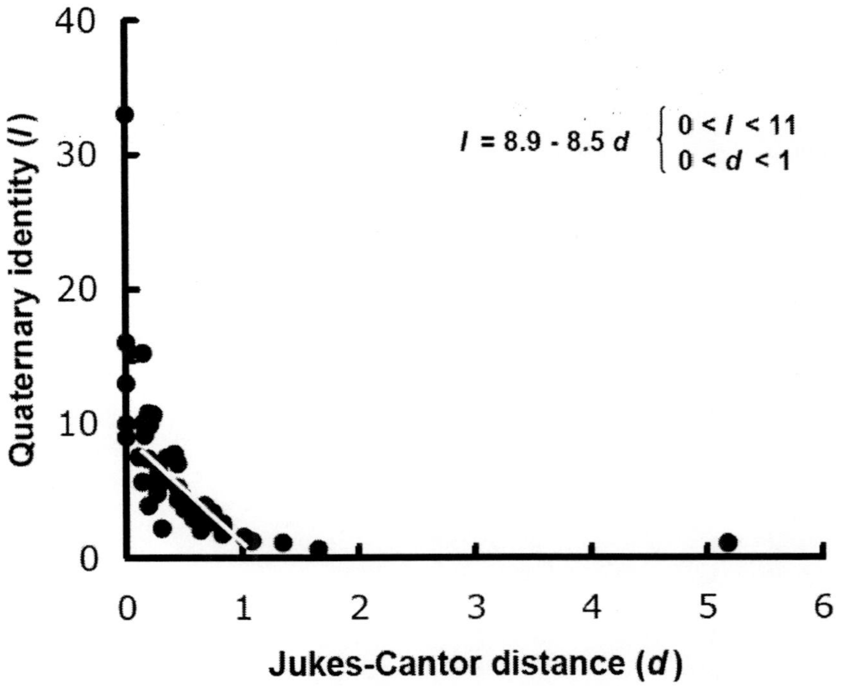

Figure 11. Depicts inverse relation between quaternary identity and Jukes-Cantor distance among LCA tRNA. Identity decreased linearly as mutation distance increased (goodness of fit, $r^2 = 0.60$), over a range encompassing thirty-seven of forty-eight (77 per cent) tRNA species. Identity (quarts) and distance values are for Asn-tRNA versus forty-eight tRNA species in table 2 and figure 4.

Results in figure 11 demonstrate, as anticipated, that quaternary identity has an inverse relation with mutation distance. A linear fall-off (slope, -8.5 quarts/unit distance) in identity (Eqn. 3) accompanies increases in Jukes-Cantor distance (Eqn. 2), over a range that included 77 per cent of the forty-eight LCA tRNA species examined.

Quaternary identity levels (Eqn. 3) helped identify tRNA ancestral to LCA tRNA species. Adaptor and ancestral tRNA were time-ordered from attached aa 'code age', inferred from path-distance. [13, 15] A tree was obtained portraying pre-LCA tRNA diversification in relation to stages identified in code evolution.

Codon Contiguity

Contiguity was statistically evaluated for base triplets encoding: (1) same-family aa, (2) same-region aa in the pre-LCA tRNA tree, and (3) same-domain aa in the genetic code. Nearest-neighbor triplet frequency between a designated aa and other members of the same set was appraised using probabilities from a hypergeometric distribution,

$$p(x) = \binom{a}{x}\binom{b}{n-x}\bigg/\binom{a+b}{n} \qquad (4)$$

where $p(0 < x \leq n)$ is the probability for chance occurrence of x nearest-neighbor triplets between a specified aa and other same-set members, when they are encoded by a total of n base triplets. A total of a triplets are nearest-neighbor to triplets of the given aa, and a total of b triplets are non-contiguous with its codons. The quantities in parenthesis are binomial coefficients. $p(x = 0)$ was defined to equal one. Three STOP signal triplets were omitted from these calculations. Codons differing solely by a 3'-pyrimidine were treated as indistinguishable. Codon-contiguity probabilities for each of the standard set of aa were combined by the method of Fisher. [37]

Amino Acids

Chemical homology among same-family aa was compared with homology achieved among different-family aa. The difference in mean Sneath coefficient [47] for intra- and inter-family homology was assessed for statistical significance, using a t-test adjusted for unequal sample size and variance. [36]

Differences in the timing of α-amine and α-carboxyl addition during aa synthesis were expressed in terms of path-length, measured from the citrate cycle, when each functional group was added in the synthesis pathway of a given aa. Goodness of fit between α-amine and α-carboxyl path-length distributions, in aa biosynthesis, was assessed with a Kolmogorov-Smirnov test. [42]

Acknowledgements

Nelida V. Davis provided helpful assistance during preparation of the manuscript.

Notation

Amino acid superscripts refer to stage of code evolution at aa addition, based on the effective number of reaction steps (from citrate cycle) in each synthesis pathway.

References

[1] Crick, FHC. (1955). A note for the RNA tie club. Reprinted by: Hoagland, MB. (1960). The relationship of nucleic and protein synthesis as revealed by studies in cell-free

systems. In: Chargaff, E, Davidson, JN, editors. *The Nucleic Acids.* New York: Academic Press; *3*, 349-408.

[2] Danchin, A. (1989). Homeotopic transformation and the origin of translation. *Prog. Biophys. Mol. Biol, 54,* 81-86.

[3] Soll, D. (1993). Transfer RNA: an RNA for all seasons. In: Gesteland, RF, Atkins, JF, editors.*The RNA World.* Plainview: Cold Spring Harbor Laboratory Press; 157-184.

[4] Marcker, K; Sanger, F. (1964). N-Formyl-methionyl-sRNA. *J. Mol. Biol, 8,* 835-840.

[5] Wilcox, M; Nirenberg, M. (1968). Transfer RNA as a cofactor coupling amino acid synthesis with that of protein. *Proc. Natl. Acad. Sci. USA*, *55,* 229-236.

[6] Schon, A; Kannangara, CG; Gough, S; Soll D. (1988). Protein biosynthesis in organelles requires misaminoacylation of tRNA. *Nature, 331,* 187-190.

[7] Forchammer, K; Bock A. (1991). Selenocysteine synthesis from *Escherichia coli* - analysis of the reaction sequence. *J. Biol. Chem*, *266,* 6324-6328.

[8] Sauerwald, A; Zhu, W; Major, TA; Roy, H; Palioura, S; Jahn, D; Whitman, WB; Yates, JR; Ibba, M; Soll, D. (2005). RNA-dependent cysteine biosynthesis in archaea. *Science*, *307*, 1969-1972.

[9] Berry-Lowe, S. (1974). The chloroplast glutamate tRNA gene for δ-aminolevulinate synthesis. *Carlsberg Res. Commun, 32*, 197-210.

[10] Yanofsky, C. (2000). Transcription attenuation: once viewed as a novel regulatory strategy. *J. Bacteriol, 182*, 1-8.

[11] Weber, U; Beier, H; Gross, HJ. (1996). Another heritage from the RNA world: self-excision of intron sequences from nuclear pre-tRNAs. *Nuc. Acids Res, 24*, 2212-2219.

[12] Dillon, LS. (1973). The origins of the genetic code. *Botanical Rev, 39*, 301-345.

[13] Davis, BK. (1999). Evolution of the genetic code. *Prog. Biophys. Mol. Biol, 72*, 157-243.

[14] Davis, BK. (2002). Molecular evolution before the origin of species. *Prog. Biophys. Mol. Biol*, *79*, 77-133.

[15] Davis, BK. (2006). Making sense of the genetic code with the path-distance model. In: Columbus, F, editor. *New Research Communications on Messenger RNA.* Hauppauge: Nova Science; 1-32.

[16] Woese, CR. (1967). *The Genetic Code.* New York: Harper and Row.

[17] Taylor, FJR; Coates, D. (1989). The code within codes. *BioSystems*, *22*, 177-187.

[18] Lim, V; Curran, P. (2001). Analysis of codon:anticodon interactions within the ribosome provides new insights into codon reading and genetic code structure. *RNA, 7*, 942-957.

[19] Dunnill, P. (1966). Triplet nucleotide-amino-acid pairing: a stereochemical basis for the division between protein and non-protein amino acid. *Nature*, *210,* 1267-1268.

[20] Perlwitz, MD; Burks, C; Waterman, MS. (1988). Pattern analysis of the genetic code. *Advan. App. Math*, *9*, 7-21.

[21] Biro, JC; Benyo, B; Sansom, C; Szlavecz, A; Fordos, G; Micsik, T; Benyo, Z. (2003). A common periodic table of codons and amino acid. *Biochem. Biophys. Res. Commun, 306*, 408-415.

[22] Brooks, DJ; Fresco, JR; Lesk, AM; Singh M. (2002). Evolution of amino acid frequencies in proteins over deep time: Inferred order of introduction of amino acids into the genetic code. *Mol. Biol. Evol*, *19*, 1645-1655.

[23] Brooks, DJ; Fresco, JR. (2003). Greater GNN pattern bias in sequence elements encoding conserved residues of ancient proteins may be an indicator of amino acid composition of early proteins. *Gene*, *303*, 177-185.

[24] Eigen, M; Lindemann, BF; Tietze, M; Winkler-Oswatitsch, R; Dress, A; von Haeseler, A. (1989). How old is the genetic code? statistical geometry of tRNA provides an answer. *Science*, *244*, 673-679.

[25] Bermudez, CJ; Daza, EE; Andrade, E. (1999). Characterization and comparison of Escherichia coli transfer RNAs based on secondary structure. *J. Theor. Biol*, *197*, 193-205.

[26] Chaley, MB; Korotkov, EV; Phoenix, DA. (1999). Relationships among isoacceptor tRNAs seem to support the coevolution theory origin of the genetic code. *J. Mol. Evol*, *48*, 168-177.

[27] Xue, H; Tong, K-L; Marck, C; Grosjean, H; Wong, JT-F. (2003). Transfer RNA paralogs: evidence for genetic code-amino acid biosynthesis coevolution and an archaeal root of life. *Gene*, *310*, 59-66.

[28] Sprinzl, M; Horn, C; Brown, M; Ioudovitch, A; Steinberg, S. (1998). Compilation of tRNA sequences and sequences of tRNA genes. *Nuç. Acids Res*, *26*, 148-153.

[29] Saks, ME; Sampson, JR. (1995). Evolution of tRNA recognition systems and tRNA gene sequences. *J. Mol. Evol*, *40*, 509-518.

[30] Greenberg, DM. (1969). Biosynthesis of amino acids and related compounds. In: Greenberg, DM, editor. *Metabolic Pathways.* New York: Academic Pres; *3*, 237-315.

[31] Rodwell, VW. (1969). Biosynthesis of amino acids and related compounds. In: Greenberg DM, editor. *Metabolic Pathways.* New York: Academic Press; *3*, 317-373.

[32] Michal, G. (1978) *Biochemical pathways*. Indianapolis: Boehringer Mannheim Biochemicals.

[33] Weiss, PM; Chen, C-Y; Cleland, WW; Cook, PF. (1988). Use of primary deuterium and ^{15}N isotope effects to deduce the relative rates of steps in the mechanisms of alanine and glutamate dehydrogenase. *Biochemistry*, *27*, 4814-4822.

[34] Saks, ME; Sampson, JR; Abelson, J. (1988). Evolution of a transfer RNA gene through a point mutation in the anticodon. *Science*, *279*, 1665-1670.

[35] Davis, BK. (2004). Expansion of the genetic code in yeast: making life more complex. *BioEssays*, *26*, 111-115.

[36] Snedecor, GW; Cochran, WG. (1967). *Statistical Methods.* Ames: Iowa State University Press.

[37] Fisher, RA. (1958). *Statistical Methods for Research Workers.* Edinburgh: Oliver and Boyd.

[38] Davis, BK. (2005). Coevolution theory of the genetic code: is the precursor-product hypothesis invalid? *BioEssays*, *27*, 1308.

[39] Saitou, N; Nei, M. (1987). The neighbor-joining method: a new method for reconstructing phylogenetic trees. *Mol. Biol. Evol*, *4,* 406-425.

[40] Jukes, TH; Cantor, CR. (1969). Evolution of proteins. In: Munro, HN, editor. *Mammalian Protein Metabolism.* New York: Academic Press; *3*, 21-132.

[41] Christophersen, S. (2004). Reconstruction and preliminary characterization of the evolutionary origin of iron-sulfur proteins: the oldest known protein and its relation to the origin of life. Thesis, s991053. Lyngby: Technical University of Denmark.

[42] Siegel, S; Castellan, NJ. (1988). *Non-Parametric Statistics for the Behavioral Sciences.* New York: McGraw-Hill.

[43] Ronneberg, TA; Landweber, LF; Freeland, S J. (2000). Testing a biosynthetic theory of the genetic code: Fact or artifact? *Proc. Natl. Acad. Sci. USA*, *97*, 13690-13695.

[44] Bretscher, MS; Goodman, HM; Menninger, JR; Smith, JD. (1965). Polypeptide chain termination using synthetic polynucleotides. *J. Mol. Biol. 14*, 634-639.

[45] Sneath, PHA. (1966). Relations between chemical structure and biological activity in peptides. *J. Theor. Biol*, *12*, 157-195.

[46] Woese, CR; Olsen, GJ; Ibba, M; Soll D. (2000). Aminoacyl-tRNA synthetases, the genetic code, and the evolutionary process. *Microbiol. Mol. Biol. Rev*, *64*, 202-236.

[47] Eriani, G; Delarue, M; Poch, O; Gangloff, J; Moras, D. (1990). Partition of aminoacyl-tRNA synthetases into two classes on the basis of two mutually exclusive sets of sequence motifs. *Nature*, *347*, 20.3-206.

[48] Walter, KU; Vamvaca, K; Hilvert D. (2005). An active enzyme constructed from a 9-amino acid alphabet. *J. Biol. Chem*, *280*, 37742-37746.

[49] Cavalcanti, ARO; Leite, ES; Neto, BB; Ferreira, R. (2004). On the classes of aminoacyl-tRNA synthetases, amino acids and the genetic code. *Origins of Life*, *34*, 407-420.

[50] Francklyn, C. (2003). tRNA synthetase paralogs: evolutionary links in the transition from tRNA-dependent amino acid biosynthesis to *de novo* biosynthesis. *Proc. Natl. Acad. Sci. USA*, *100*, 9650-9652.

[51] Wetzel, R. (1995). Evolution of the aminoacyl-tRNA synthetase and the origin of the genetic code. *J. Mol. Evol*, *40*, 545-550.

[52] Davis, BK. (2006). Imprint of tRNA diversification on the genetic code. http://www.archive.org/details/tRNA_imprint_on_code

[53] Kinchin, AI. (1957). *Mathematical Foundations of Information Theory.* Silverman, RA, Friedman, MD, translators. New York: Dover.

APPENDIX A

Distribution of Amino Acid Identity, Core Group and Anticodon Bases in LCA tRNA Tree

Data relating to the distribution of aa identity, core group and anticodon 3'- and mid-base, in cluster-regions identified in the LCA tRNA tree (figure 4), are presented in tables 1A to 4A. Statistical analysis of the data is provided and its significance discussed in this chapter (see Channeled diversification of tRNA species in the pre-LCA era).

Table 1A. Distribution of adaptors for same-family amino acids in LCA tRNA tree regions[1]

Tree region	Amino acid family	Total tRNA (no.)	tRNA for aa family (no.)	Frequency specific tRNA	Probability[2] p	-2 ln p
1	Glu	9	7	0.167	$8.942x10^{-5}$	18.644
2	Pyr	8	6	0.229	$2.412x10^{-3}$	12.055
3	Asp	18	14	0.375	$5.078x10^{-4}$	15.171
4	Ser	13	7	0.188	$4.022x10^{-3}$	11.032
5	Shi	0	0	0.042	1.0	0
	Total:	48	34		χ^2(df = 10) :	56.902
				Combined probability:		$1.391x10^{-8}$

[1] Data are from tRNA tree in figure 4.

[2] Binomial probability (Eqn. 1) for number of specific aa adaptors in designated LCA tRNA tree region. Probabilities were combined by Fisher's method. [37]

Table 2A. Distribution of tRNA types and subtypes between regions in LCA tRNA tree[1]

Tree region	Type/ subtype of tRNA	Total tRNA (no.)	Speci fic tRNA (no.)	Frequency specific tRNA	Probability[2] p	-2 ln p
1	ID	9	8	0.250	$1.030x10^{-4}$	18.363
2	IA	8	7	0.521	$3.984x10^{-2}$	6.446
3	IA	18	15	0.521	$5.050x10^{-3}$	10.577
4	II	13	7	0.188	$4.022x10^{-3}$	11.032
5	IB	0	0	0.042	1.0	0
	Total:	48	37		χ^2(df = 10) :	46.416
				Combined probability:		$1.206x10^{-6}$

[1] Data are from tRNA tree in figure 4.

[2] Binomial probability (Eqn. 1) for number of specific kinds of tRNA in designated tree region.

Table 3A. Distribution of tRNA with specific 3'-anticodon base in regions of LCA tRNA tree[1]

Tree region	anticodon 3'-base (nt. 36)	Total tRNA (no.)	tRNA with given nt.36 (no.)	Frequency tRNA with nt. 36	Probability[2] p	-2 ln p
1	G	9	6	0.271	1.284×10^{-2}	8.710
2	C	8	7	0.208	1.078×10^{-4}	18.271
3	U	18	12	0.292	8.896×10^{-4}	14.049
4	A	13	5	0.188	5.665×10^{-2}	5.742
5	A	0	0	0.188	1.0	0
	Total:	48	37		χ^2(df = 10):	46.772
				Combined probability:		1.039×10^{-6}

[1] Data are from tRNA tree in figure 4.
[2] Binomial probability (Eqn. 1) for number of tRNA with specific 3'-anticodon base (nt. 36) in tree region.

Table 4A. Distribution of tRNA with specific mid-anticodon base in regions of LCA tRNA tree[1]

Tree region	anticodon mid-base (nt. 35)	tRNA in region (no.)	tRNA with given nt.35 (no.)	Frequency tRNA with nt.35	Probability[2] p	-2 ln p
1	U	9	5	0.208	7.212×10^{-2}	5.259
2	A	8	4	0.271	0.106	4.480
3	A	18	5	0.271	0.206	3.163
4	C	13	6	0.250	5.592×10^{-2}	5.768
5	-	0	0	-	1.0	0
	Total:	48	37		$\chi 2$(df = 10):	18.670
				Combined probability:		4.466×10^{-2}

[1] Data are from tRNA tree in figure 4.
[2] Binomial probability (Eqn. 1) for number of tRNA with specific mid-anticodon base (nt. 35) in tree region.

In: Messenger RNA Research Perspectives
Editor: Takeo Takeyama, pp. 81-94
ISBN: 978-1-60021-814-9

Chapter II

Expression of Ammonia Monooxygenase Gene in Domestic Wastewater Treatment Process

Yoshitaka Ebie[1], Satoshi Tsuneda[2] and Yuhei Inamori[1]

[1] Research Center for Material Cycles and Waste Management, National Institute for Environmental Studies, 16-2 Onogawa, Tsukuba, Ibaraki 305-8506, Japan

[2] Department of Chemical Engineering, Waseda University, 3-4-1 Ohkubo, Shinjuku, Tokyo 169-8555, Japan

Abstract

RT-PCR targeting *amoA* mRNA which coding a subunit of ammonia monooxygenase (AMO) was combined with fingerprinting methods to elucidate effect of some important environmental conditions on transcriptional activity of *amoA* gene. Expression of *amoA* gene in domestic wastewater treatment process was monitored by combination analysis of RT-PCR and DGGE. Sludge samples were collected from bench-scale reactors. After total RNA was extracted, RT-PCR targeting *amoA* mRNA was performed. DGGE analysis was used to separate the amplified products according to their sequence and thereby to visualize individual community members. Ammonia oxidizers that were demonstrated by the presence of *amoA* mRNA were regarded as metabolically active ones. The DGGE profiles of *amoA* mRNA RT-PCR products obtained from the samples collected at the initial stage of the reactor operation showed decrease of the diversity of *amoA* mRNA. When influent load was changed with reduction of hydraulic retention time (HRT) or increase of concentration of wastewater, new DGGE bands newly appeared on only the short HRT condition. The *amoA* sequences of DGGE bands affiliated to the *Nitrosomonas eutropha* lineage disappeared during the beginning of the operation. On the other hand, all the *amoA* sequences of new DGGE bands affiliated to the genus *Nitrosomonas*, and formed a phylogenetic lineage including no cultured representatives. Effect of ammonium concentration and temperature on transcriptional activity of *amoA* gene was also investigated. Activated sludge was collected from artificial domestic wastewater treatment reactor and incubated with different ammonium concentration (25, 100 and 250 mg l^{-1}). The incubation temperature was adjusted to 20 or 30°C. Ammonium consumption rate during the first few hours was higher at higher temperature and lower initial ammonium concentration. The T-RFLP profiles of *amoA* mRNA showed different effect of ammonium concentration and temperature on transcriptional activity of *amoA* for Genus *Nitrosomonas* and *Nitrosospira.* The observed main peak in the all samples was of the length associated with *Nitrosomonas*-like *amoA* (219 bp). Other peaks of *Nitrosomonas*-like *amoA* (491 bp) and *Nitrosospira*-like *amoA* gene (283 bp) were hardly detected in the samples incubated at 30°C. These minor peaks were observed in the samples incubated at 20°C and the height of these peaks were relatively high in the samples incubated with high initial ammonium concentration. Cloning analysis revealed that the dominant sequence of expressed *amoA* mRNA is not closely affiliated with that of any isolates. These results of mRNA analysis suggest that uncultured but active *Nitrosomonas* strains dominated in high ammonium concentration and high temperature condition.

INTRODUCTION

The process of nitrification, defined as the oxidation of ammonia to nitrate via nitrite, is of considerable importance not only in the terrestrial nitrogen cycle but also in the biological wastewater treatment process. Molecular techniques such as denaturing gradient gel electrophoresis (DGGE) and fluorescence in situ hybridization (FISH) based on SSU rDNA, rRNA and functional genes have been developed to analyze populations of ammonia oxidizer [1-3]. However, molecular analyses based on 16S rDNA information and the gene amoA are unable to estimate the actual in situ ammonia-oxidizing activity. In fact, Mendum et al. reported that populations of ammonia oxidizer did not change significantly, even with a large increase in nitrification rates following the addition of ammonium to soils [4]. Techniques that focus on specific and transient markers of in situ metabolism, such as mRNA, offer the possibility of measuring specific metabolic activity. The retrieved transcripts can be used to compare the environmental expression of individual members of gene families. Thus, when applied to field samples, mRNA-based analysis may be useful in determining the relationships between environmental conditions in microbial habitats and particular in situ activities of native microorganisms [5].

In this chapter, we introduce an approach to investigating the community structure of ammonia oxidizers, which are responsible for ammonia oxidation, based on the expression of amoA mRNA.

MICROBIAL COMMUNITY AND ACTIVITY IN NITRIFICATION PROCESS

The biological removal of nitrogen compounds is an integral part of most modern wastewater treatment facilities to preserve water environmental resources. In such wastewater treatment facilities, nitrification is a key step in reducing the ammonia content of domestic and industrial wastewaters. Two distinct, physiologically defined groups of bacteria catalyze the two separate steps involved in nitrification [6, 7]. First, chemolithoautotrophic ammonia oxidizers convert ammonia to nitrite, which is subsequently transformed to nitrate by nitrite oxidizers. Both oxidation processes are prerequisite for biological nitrogen removal from wastewater via subsequent denitrification, however, ammonia oxidation is a rate-limiting step in biological nitrification due to the low growth rates of ammonia oxidizers and their sensitivity to toxic shocks, pH and temperature swings [7, 8]. Therefore, it is indispensable to estimate the structure of the microbial community and the physiological activity of ammonia oxidizers in activated sludge or biofilms. Traditional cultivation methods such as the most-probable-number technique or selective plating are very time-consuming, have low efficiencies and lead to a collection of ammonia oxidizers that are not representative of their actual diversity and abundance in the habitats [9, 10]. Therefore, molecular analyses such as fluorescence in situ hybridization (FISH) and denaturing gradient gel electrophoresis (DGGE) based on 16S rDNA (rRNA) and functional gene, which do not require cultivation, appear to be promising alternatives [11, 12]. Ammonia oxidizers in environmental samples were first detected by Sinigalliano et al. [13] using PCR primers to specifically amplify the functional gene encoding the α-subunit of ammonia monooxygenase (amoA). Since then, ammonia oxidizer populations have been studied by many researchers targeting the amoA gene [14-17]. However, gene-based molecular analyses are inadequate to demonstrate the actual in situ physiological activity. Therefore, deterioration of nitrogen removal performance that is occasionally observed in the wastewater treatment process could not be understood by the analysis only based on 16S rRNA gene and/or the amoA gene.

Recently, mRNA-based analysis called transcriptome has become center of attraction because mRNA transcripts are short-lived and its sequences are enough unique to identify with high specificity. When nucleic acid sequences information of specific transcripts is available for the analysis of a microbial community, this approach obviates the need for culturing microorganisms or incubating environmental samples, which eliminates the biases and limitations frequently observed in other methods [18]. Therefore, mRNA-based analysis may be one of the alternatives in determining in situ functional activities of microorganisms.

Combination Analysis of RT-PCR and DGGE to Monitor Expression of Ammonia Monooxygenase Gene

In this study, reverse transcription (RT) and PCR were combined with DGGE (Denaturing Gradient Gel Electrophoresis) to detect mRNA concerning ammonia oxidation and to elucidate community structure of ammonia oxidizers based on in situ ammonia-oxidizing activity. Especially, we focused on AMO (ammonia monooxygenase) which catalyzes the first step of ammonia oxidation, oxidation of ammonia to hydroxylamine, and expression of amoA gene which coding a subunit of AMO in domestic wastewater treatment process was monitored by this RT-PCR-DGGE method.

Materials and Methods

Two aerobic-anaerobic circulation wastewater treatment reactors were operated at 20°C. Artificial domestic wastewater containing 45 mg l^{-1} nitrogen was used as influent of these bench-scale reactors. Nitrogen load on the aerobic tank was set at 0.073 kg-N m^{-3} day^{-1} at the beginning of the operation (period 1). In period 2, nitrogen load in both runs was increased to 0.13 kg-N m^{-3} day^{-1} by the different two ways. Influent water volume was increased and HRT (hydraulic retention time) was changed from 2.5 to 1.4 days in Run 1. In contrast, concentration of the influent artificial wastewater was increased and nitrogen concentration was changed from 45 to 79 mg L^{-1} in Run 2. Sludge samples were collected from these reactors. Extraction of total RNA from the sludge samples was performed with an ISOGEN reagent (Nippon Gene Co., Tokyo, Japan), according to the manufacture's protocol except cell disruption procedure [19]. Homogenization was performed with Mini-beadbeater-8 (BioSpec Products, Bartlesville, Okla.) on the "HOMOGENIZE" setting for 2 min in the presence of 1 g of glass beads (diameter, 0.1 mm; BioSpec Products). Remaining DNA was digested by RNase-free DNase I (TaKaRa Shuzou Co. Ltd., Siga, Japan) according to a procedure of Ward et al. [20]. Extracted RNA samples were stored at -80°C.

RT-PCR was performed with ReverTra Ace (TOYOBO) and random primer (TOYOBO). The reverse transcription reaction was carried out at 42°C for 50 min, followed by RNase inactivation at 99°C for 5 min and final cooling at 4°C for 5 min with GeneAmp PCR System 9700 (Applied Biosystems, Foster City, Calif.). The first-strand cDNA transcribed from total RNA was amplified with primers amoA-1F and amoA-2R [21]. The PCR reaction was performed with KOD Plus (TOYOBO) [19]. PCR of extracted RNA without the RT step was carried out as a control for DNA contamination. For DGGE, RT-PCR amplification products were used as template for a second PCR. The latter PCR was done with primer amoA-1F-GC that was attached to the 40-nucleotide GC-rich sequence [19]. The amoA-2R primer has two degeneracies and primers with no degeneracy were constructed [22] for second PCR.

DGGE was performed with a D-Code Multi-system (Bio-Rad Laboratories, Hercules, Calif.) to separate the amplified products according to their sequence and thereby to visualize individual community members. Major DGGE bands were excised from the gel. DNA was extracted from the excised gels and sequenced with the amoA-1F and amoA-2R primer.

Community Members Demonstrated by RT-PCR-DGGE

Nitrification efficiency during startup of the operation (period 1) and stationary phase (period 2) was shown in figure 1. Nitrification efficiency rose to 95% by 30 days operation. The DGGE profiles of amoA mRNA RT-PCR products obtained in period 1 showed decrease in diversity of amoA mRNA (figure2). Ammonia oxidizers that were demonstrated by the presence of amoA mRNA were regarded as metabolically active ones. The bands obtained could be categorized into three: detected only in the samples collected at the beginning of the operation (DGGE-bands 4 and 5), became clear at stationary phase of the operation (DGGE-band 3), and observed through the operation (DGGE-bands 1 and 2).

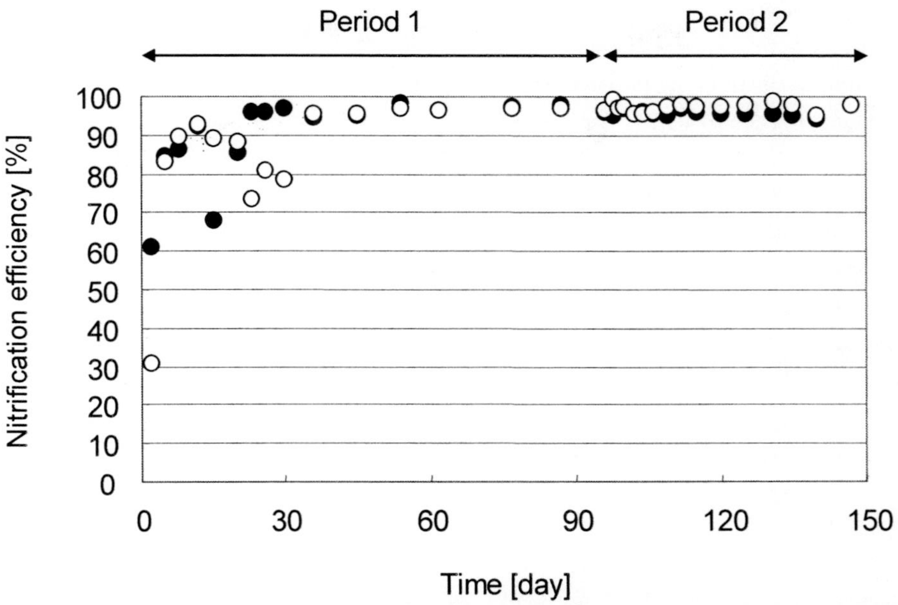

Figure 1. Nitrification efficiency in the lab-scale artificial domestic wastewater treatment process. Symbols: closed circle, Run 1; open circle, Run 2.

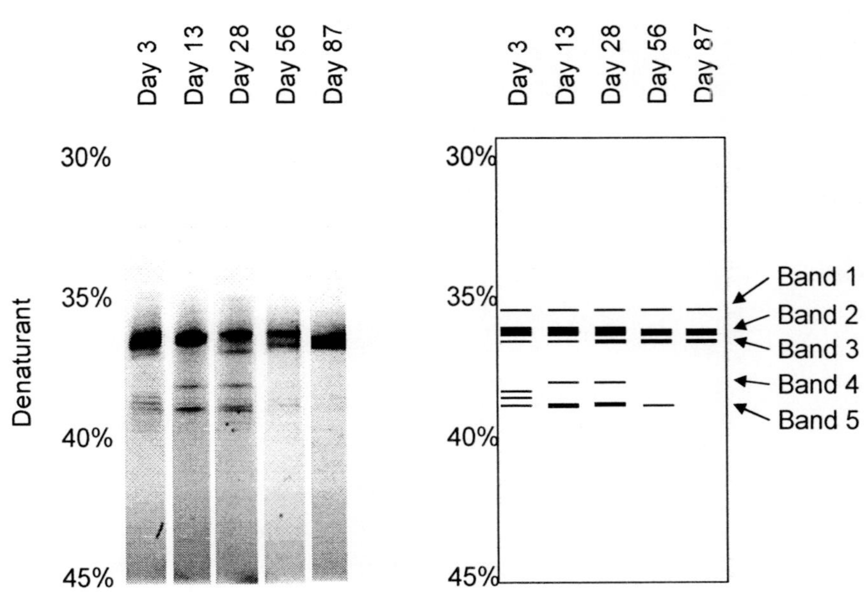

Figure 2. DGGE band profiles of the *amoA* mRNA RT-PCR products obtained from Run 1 in period 1. DGGE bands pointed by arrows were sequenced for phylogenetic analysis.

Decrease of the diversity of amoA mRNA indicates that seeding sludge collected from wastewater treatment plant treating actual domestic wastewater harbored a lot of kinds of ammonia oxidizers and simple artificial wastewater used in this study led to simple community of ammonia oxidizers. Sequencing of the bands revealed that DGGE-bands 4 and 5 are similar to Nitrosomonas eutropha. This ammonia oxidizer prefers high ammonium loading rate and is detected in many wastewater treatment plants [15]. Although ammonium loading rate of this reactor (0.073 kg-N m^{-3} day^{-1}) was not so high, ammonium loading rate per cell must be high when the quantity of ammonia oxidizers was low at the beginning of the operation. This might be a reason for the domination of this kind of ammonia oxidizer. In fact, these bands disappeared at the stationary phase. These results showed that members of active ammonia oxidizers were changed with changing of influent wastewater and number of the bacteria in the reactor, and RT-PCR-DGGE method based on amoA mRNA could demonstrate this community succession of bacteria expressing ammonia-oxidation activity.

Effects of Ammonium Concentration and HRT on Ammonia Oxidizers

At stationary phase of the reactor operation, ammonium loading rate was increased to 0.13 kg-N m^3 day^{-1} (period 2). Influent water volume was increased and HRT shortened in Run 1, whereas concentration of the influent artificial wastewater was increased in Run 2. In this period, nitrification efficiency was kept at over 90% even though nitrogen load increased. As a result of amoA mRNA monitoring, two clear DGGE bands appeared in Run 1 (figure 3). This means existence of bacteria increasing amoA mRNA transcript in response to changes in environmental condition. Although the same DGGE bands were observed in Run 2, it was unclear and obscured compared with that of Run 1. There were no differences in operational conditions between the samples except for the way of increase of ammonia loading rate. These results suggest that amoA mRNA-based analysis revealed different active ammonia oxidizers in Runs 1 and 2, even ammonium loading rate was the same.

Metabolically Active Ammonia Oxidizers Detected by RT-PCR-DGGE

The sequences determined in this study were compared with other available sequences by using BLAST search, and were aligned by Clustal W with DNA Data Bank of Japan (DDBJ). Neighbor-joining tree was constructed by Clustal W with DDBJ and Tree View [23]. Figure 4 shows that all ammonia oxidizers detected in the bench-scale reactor affiliated to the genus Nitrosomonas. The result of mRNA-based analysis revealed metabolically active members are certainly genus Nitrosomonas. So far, some amoA gene-based analyses have revealed that the majority of amoA sequences derived from activated sludge in sewage treatment plants are those of bacteria in the genus Nitrosomonas [3, 15]. However, DNA-based analyses detect microbes present within an environmental sample, including both dormant and inactive microbes. Analysis of specific genes can estimate the potential of these microbes but not their actual in situ activity. Conversely, mRNA-based analysis, as described in this paper, can enable differentiation of metabolically active strains from inactive strains.

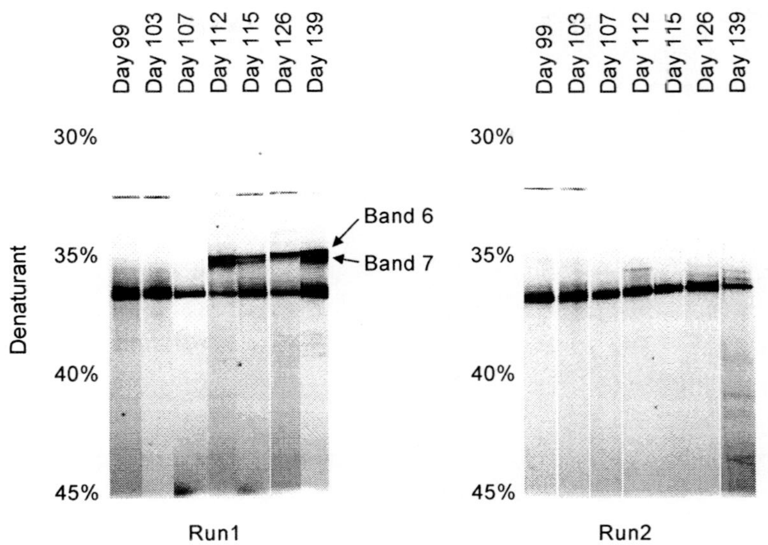

Figure 3. DGGE band profiles of the *amoA* mRNA RT-PCR products in period 2. DGGE bands pointed by arrows were applied to phylogenetic analysis.

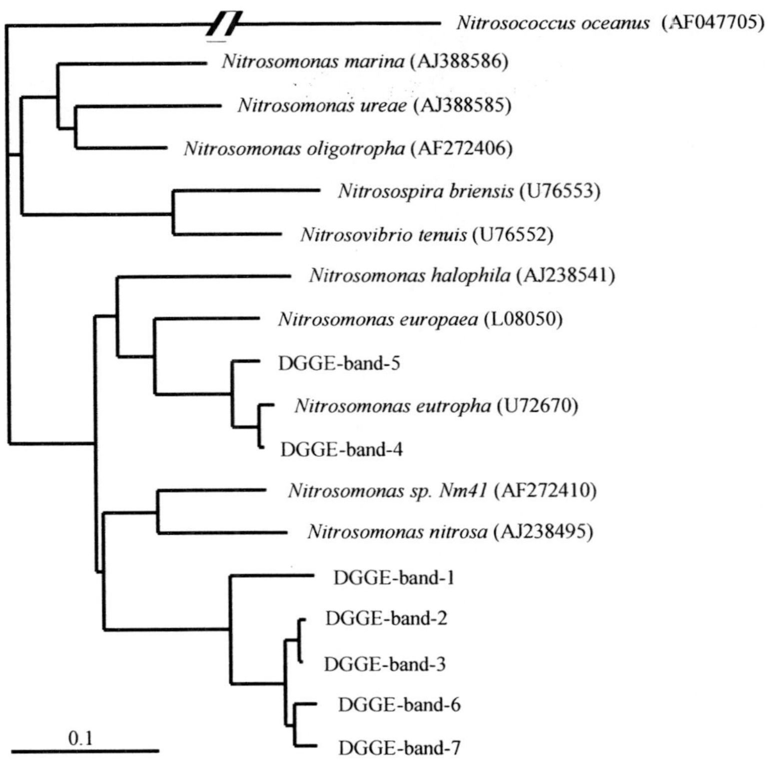

Figure 4. Neighbor-joining tree showing the relationships partial nucleotide sequences from DGGE band of *amoA* mRNA RT-PCR products with reference sequences in the database. Nucleotide sequence database accession numbers of type strains are shown in the parentheses.

DGGE-bands 4 and 5 appeared only at the beginning of the operation were similar to Nitrosomonas eutropha which are often detected in many wastewater treatment plants [15] and recognized as bacteria working under the high nitrogen load condition. On the other hand, all the other amoA sequences, DGGE-band 3 which became clear at stationary phase of the operation, DGGE-bands 1 and 2 which observed through the operation, and DGGE-bands 6 and 7 which appeared after reduction of HRT, affiliated to the genus Nitrosomonas and formed a phylogenetic lineage including no cultured representatives. These ammonia oxidizers detected by the mRNA-based analysis must responsible for ammonia oxidation; however, similarity of the amoA sequence was obviously different from any isolated bacterial amoA gene sequences. It was found that ammonia oxidizers which are not yet isolated should play an important role for ammonia oxidation.

The results obtained in this study showed low diversity comparing previous study based on amoA gene [3, 13-14, 16-17] and 16S rRNA gene [15, 26]. This low diversity might be explained by following reason. Because the amount of amoA mRNA transcript quickly changes with environmental conditions, the difference in copy number of individual amoA mRNA derived from different species widened. Thus, only major amoA mRNA sequences were detected by DGGE and resulted in the low diversity of ammonia oxidizers.

In this study, amoA mRNA was monitored as an indicator of ammonia-oxidizing activity. This mRNA-based analysis revealed microbial community actually responsible for ammonia oxidation in wastewater treatment process. Wewer et al. and Nogales et al. investigated expression of [NiFe] hydrogenase gene and nirS gene, respectively [24, 25]. They reported that community structure analyzed by mRNA-based analysis was different from that of DNA-based analysis. Therefore, comparative analysis of mRNA and its gene is one of the promising tools for further understanding of microbiological ecology in the environment [19].

T-RFLP Analysis of Expression of Ammonia Monooxygenase Gene in Different Ammonium Concentration and Temperature

Ammonia-oxidizing activity is affected by ammonium concentration and temperature. Effect of these factors on ammonia oxidation rate can be demonstrated by ammonia oxidation batch experiment. Characteristics of isolates of ammonia oxidizer to ammonium concentration and temperature can be also investigated. However, it is impossible to estimate the effect of these factors on ammonia-oxidizing activity in each ammonia oxidizer which has not been isolated.

In this study, RT-PCR targeting amoA mRNA was combined with T-RFLP (terminal restriction fragment polymorphism) to investigate effect of ammonium concentration and temperature on transcriptional activity of amoA gene.

Materials and Methods

Activated sludge was collected from artificial domestic wastewater treatment reactor and incubated with different ammonium concentration (25, 100 and 250 mg l^{-1}). The ammonia oxidation batch experiment was conducted for 4 h at 20 and 30°C, respectively. The samples were collected several times from each bottle.

Total RNA was extracted from the sludge samples with ISOGEN (Nippon Gene). Reverse transcription was done with random primer, and the cDNAs were applied to cloning analysis. Primers amoA-1F and amoA-2R that are specific for amoA of beta-Proteobacterial ammonia oxidizers were used for PCR of the cDNA derived from amoA mRNA. The cDNAs were also used as a template for T-RFLP analysis. The cDNA was amplified with 6-FAM labeled amoA-1F and amoA-2R primers. Fluorescently labeled PCR products were purified and cleaved for 4 h at 65°C with tetrameric restricted enzyme TaqI. Aliquots of digest were purified and analyzed by electrophoresis on automated sequence analyzer in GeneScan mode. The relative abundance of terminal fragments was determined by calculating the ratio between the peak height of each peak and the total peak height of all the peaks within one sample. Lengths of predominant T-RFs were theoretically compared with database in DDBJ and the clone library generated in this study.

Effect of Ammonium Concentration and Temperature on amoA Gene Expression

Batch experiment was conducted to demonstrate effect of ammonia concentration and temperature on ammonia oxidation. Nitrification rate was influenced by initial ammonium concentration and temperature. The rate of increase in nitrite and nitrate nitrogen (NO_{2+3}-N) during the first few hours was higher at higher temperature and lower initial ammonium concentration (figure 5).

Effect of ammonia concentration and temperature on ammonia-oxidizing activity in each ammonia oxidizer was investigated by use of TaqI-based T-RFLP analysis of amoA gene expression. The mRNA-based analysis is considered as an effective method to analyze ammonia oxidizers in the batch experiment. The T-RFLP profiles of amoA mRNA showed different effect of ammonium concentration and temperature on transcriptional activity of amoA for Genus Nitrosomonas and Nitrosospira (figure 6). The observed main peak in the all samples was of the length associated with Nitrosomonas-like amoA (219 bp). This means major ammonia oxidizer actually responsible to ammonia oxidation in this batch test was Genus Nitrosomonas. Furthermore, the occupation ratio of the 219 bp T-RF which represents Nitrosomonas-like amoA increased 30% in the condition of higher ammonium concentration.

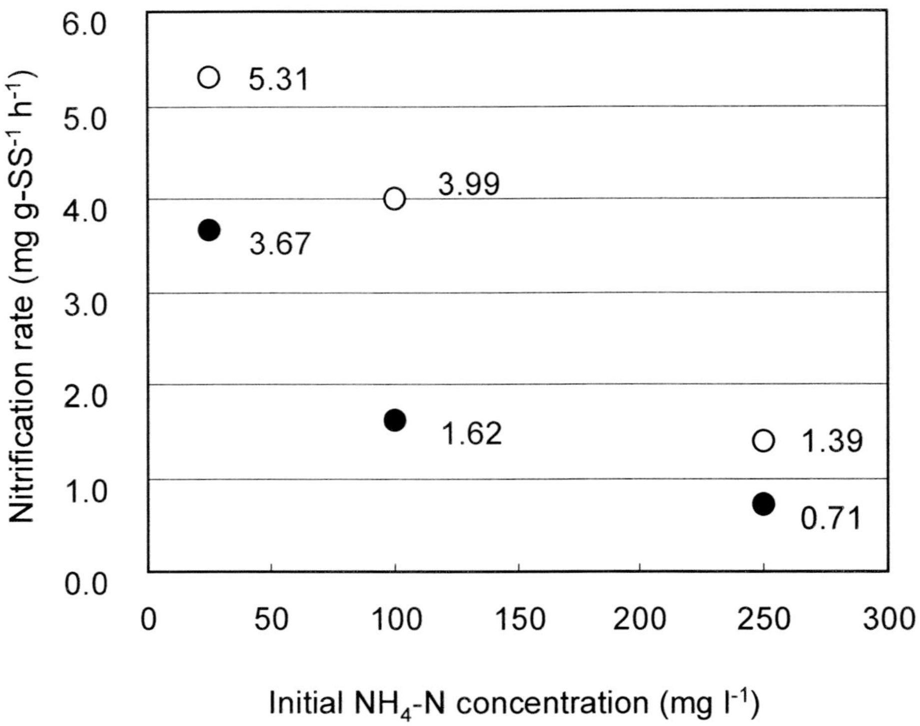

Figure 5. The Nitrification rate in different initial NH_4-N concentration and temperature. Symbols: closed circle, 20°C; open circle, 30°C.

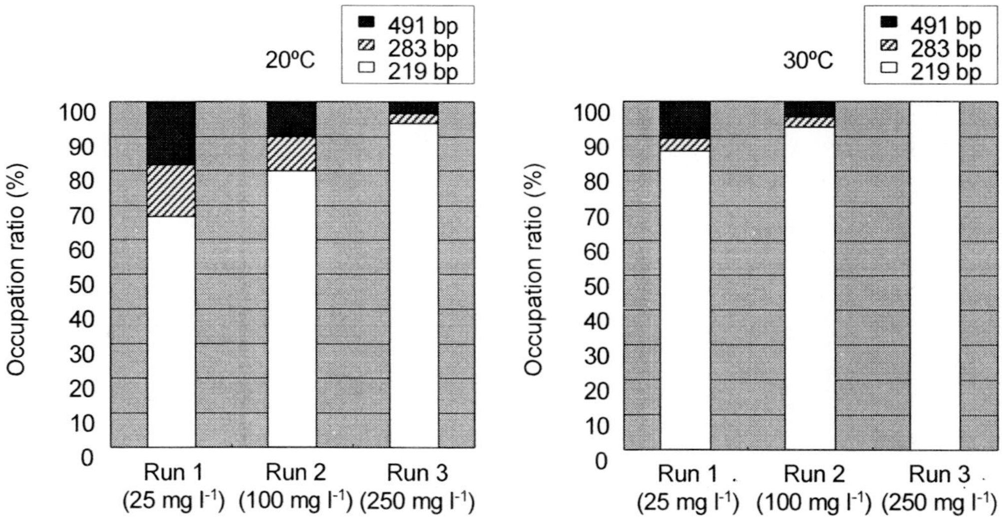

Figure 6. The occupation ratio of three major T-RFs after 120 min. The fluorescently labeled RT-PCR products of *amoA* mRNA were digested with *Taq*I.

Other peaks of Nitrosomonas-like amoA (491 bp) and Nitrosospira-like amoA gene (283 bp) were hardly detected in the samples incubated at 30°C. These minor peaks were observed in the samples incubated at 20°C and the height of these peaks were relatively high in the samples incubated with low initial ammonium concentration.

Figure 7 shows phylogenetic dendrogram constructed for partial amoA mRNA sequences. The numbers at right side indicate the size of T-RFs digested by TaqI. The dendrogram revealed that the dominant sequences of expressed amoA mRNA in this study were not similar to any type strain. Therefore, major bacteria actually responsible for ammonia oxidation might be not type strains but unisolated bacteria. These results of amoA mRNA-based analysis suggest that unisolated but active Nitrosomonas should dominate in the condition of high ammonium concentration and high temperature.

As shown in this chapter, amoA mRNA-based analysis could provide valuable information on the microbial community responsible for ammonia oxidation in not only wastewater treatment processes but also natural environments and bioremediation sites.

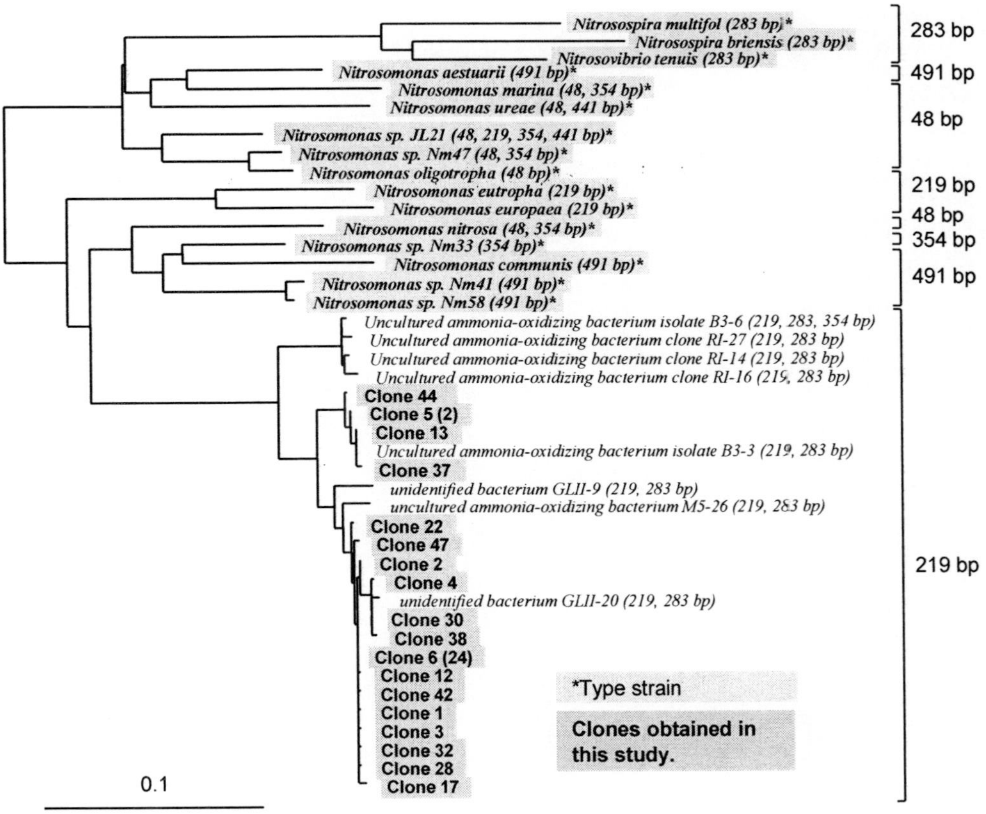

Figure 7. Phylogenetic dendrogram constructed for partial *amoA* mRNA sequences based on a neighbor-joining analysis. The numbers in square bracket showed number of clones. The numbers in parenthesis indicate the size of T-RFs which occur in *Taq*I-based T-RFLP analysis.

REFERENCES

[1] Kowalchuk GA; Stephen JR; De Boer W; Prosser JI; Embley TM; Woldendorp JW. (1997) Analysis of ammonia-oxidizing bacteria of the ß subdivision of the class *Proteobacteria* in coastal sand dunes by denaturing gradient gel electrophoresis and sequencing of PCR-amplified 16S ribosomal DNA fragments. *Appl. Environ. Microbiol.,* 63, 1489-1497.

[2] Mobarry BK; Wagner M; Urbain V; Rittman BE; Stahl DA. (1996) Phylogenetic probes for analyzing abundance and spatial organization of nitrifying bacteria, *Appl. Environ. Microbiol.,* 62, 2156-2162.

[3] Rotthauwe JH; Witzel KP; Liesack W. (1997) The ammonia monooxygenase structural gene *amoA* as a functional marker: molecular fine-scale analysis of natural ammonia-oxidizing populations. *Appl. Environ. Microbiol.,* 63, 4704-4712.

[4] Mendum TA; Sockett RE; Hirsch PR. (1999) Use of molecular and isotopic techniques to monitor the response of autotrophic ammonia-oxidizing populations of the ß subdivision of the class *Proteobacteria* in arable soils to nitrogen fertilizer. *Appl. Environ. Microbiol.,* 65, 4155-4162.

[5] Wilson MS; Bakermans C; Madsen EL. (1999) *In situ,* real-time catabolic gene expression: extraction and characterization of naphthalene dioxygenase mRNA transcripts groundwater. *Appl. Environ. Microbiol.,* 65, 80-87.

[6] Prosser, JI. (1989) Autotrophic nitrification in bacteria. *Adv. Microb. Physiol.,* 30, 125-181.

[7] Watson, SW; E. Bock; H. Harms; H-P Koops; AB Hooper. (1989) Nitrifying bacteria. In JT Staley; MP Bryant; N. Pfennig, and JG Holt (Eds.), *Bergey's manual of systematic bacteriology,* vol.3 (1808-1834). Baltimore, Md. The Williams and Wilkins Co.

[8] Princic, A; I. Mahne; F. Megusar; E. A. Paul; JM. Tiedje. (1998) Effects of pH and ammonium concentrations on the community structure of nitrifying bacteria from wastewater, *Appl. Environ. Microbiol.,* 64, 3584-3590.

[9] Phillips, CJ; Smith, Z; Embley, TM; Prosser, JI. (1999) Phylogenetic differences between particle-associated and planktonic ammonia-oxidizing bacteria of the ß-subdivision of the class *Proteobacteria* in the Northwestern Mediterranean Sea. *Appl. Environ. Microbiol.,* 65, 779-786.

[10] Wagner, M; Amman, R; Lemmer, H; Schleifer, KH. (1993) Probing activated sludge with *Proteobacteria*-specific oligonucleotides: inadequacy of culture-dependent methods for describing microbial community structure. *Appl. Environ. Microbiol.,* 59, 1520-1525.

[11] Amann, RI; W. Ludwig; K-H Schleifer. (1995) Phylogenetic identification and in situ detection of individual microbial cells without cultivation. *Microbiol. Rev.,* 59, 143-169.

[12] Muyzer, G; EC de Waal; AG Uitterlinden. (1993) Profiling of complex microbial populations by denaturing gradient gel electrophoresis analysis of polymerase chain reaction-amplified genes coding for 16S rRNA. *Appl. Environ. Microbiol.,* 59, 695-700.

[13] Sinigalliano, CD; DN Kuhn; RD Jones. (1995) Amplification of the *amoA* gene from diverse species of ammonium-oxidizing bacteria and from an indigenous bacterial population from seawater. *Appl. Environ. Microbiol.,* 61, 2702-2706.

[14] Hastings, R.C; MT Ceccherini; N. Miclaus; JR Saunders; M. Bazzicalupo; AJ McCathy. (1997) Direct molecular biological analysis of ammonia oxidizing bacteria populations in cultivated soil plots treated with swine manure. *FEMS Microbiol. Ecol.,* 23, 55-63.

[15] Purkhold, U; AP-Röser; S. Juretschko; MC Schmid; H-P Koops; M. Wagner. (2000) Phylogeny of all recognized species of ammonia oxidizer based on comparative 16S rRNA and *amoA* sequence analysis: implications for molecular diversity surveys. *Appl. Environ. Microbiol.,* 66, 5368-5382.

[16] Stephen, JR; Y-J Chang; SJ Macnaughton; GA Kowalchuk; KT Leung; CA Flemming; DC White. (1999) Effect of toxic metals on indigenous soil ß-subgroup Proteobacterium ammonia oxidizer community structure and protection against toxicity by inoculated metal-resistant bacteria. *Appl. Environ. Microbiol.,* 65, 95-101.

[17] Sakano, Y; L Kerkhof. (1998) Assessment of changes in microbial community structure during operation of an ammonia biofilter with molecular tools. *Appl. Environ. Microbiol.,* 64, 4877-4882.

[18] Madsen, EL. (1998) Epistemology of environmental microbiology. *Environ. Aci. Technol.,* 32, 429-439.

[19] Ebie Y; Noda N; Miura H; Matsumura M; Tsuneda S; Hirata A; Inamori Y. (2004) Comparative analysis of genetic diversity and expression of amoA gene in wastewater treatment process, *Appl. Microbiol. Biotechnol.,* 64, 740-744.

[20] Ward DM; Roberts ALR; Weller R. (1995) 1.2.3. Methods for extracting RNA or ribosomes from microbial mats and cultivated microorganisms. In Akkermans ADL, Elsas JDV, de Bruijn FJ (Eds.), *Molecular Microbial Ecology Manual.* (1-14). Netherlands, Kluwer Academic Publishers.

[21] Rotthauwe, JH; Witzel, KP; Liesack, W. (1997) The ammonia monooxygenase structural gene amoA as a functional marker: molecular fine-scale analysis of natural ammonia-oxidizing populations, *Appl. Environ. Microbiol.,* 63, 4704-4712.

[22] Nicolaisen, MH; Ramsing, NB. (2002) Denaturing gradient gel electrophoresis (DGGE) approaches to study the diversity of ammonia-oxidizing bacteria. *J. Microbiol. Methods,* 50, 189-203.

[23] Thompson, JD; Higgins, DG; Gibson, TJ. (1994) Clustal W: Improving the sensitivity of progressive multiple sequence alignment through sequence weighting position-specific gap penalties and weight matrix choice. *Nucleic Acids Res.,* 22, 4673-4680.

[24] Wawer, C; Jetten, MSM; Muyzer, G. (1997) Genetic diversity and expression of the [NiFe] hydrogenase large-subunit gene of Desulfovibrio spp. in environmental samples, *Appl. Environ. Microbiol.,* 63, 4360-4369.

[25] Nogales, B; Timmis, KN; Nedwell, DB; Osborn, AM. (2002) Detection and diversity of expressed denitrification genes in estuarine sediments after reverse transcription-PCR amplification from mRNA, *Appl. Environ. Microbiol.,* 68, 5017-5025.

[26] Kowalchuk, GA; Naoumenko, ZS; Derikx, PJL; Felske, A; Stephen, JR; Arkhipchenko, IA. (1999) Molecular analysis of ammonia-oxidizing bacteria of the ß-subdivision of

the class Proteobacteria in compost and composting materials, *Appl. Environ. Microbiol.*, 65, 396-405.

In: Messenger RNA Research Perspectives
Editor: Takeo Takeyama, pp. 95-113

ISBN: 978-1-60021-814-9

Chapter III

Phase- and Tissue-Specific Heterogeneity in the Expression of Genes Encoding Components of Proteolytic Systems During Fasting, Refeeding, and after Leptin Treatment

Fabrice Bertile*[*], *Sandrine Zahn and Thierry Raclot
Institut Pluridisciplinaire Hubert Curien, Département « Ecologie, Physiologie et Ethologie », CNRS-ULP UMR 7178, 23 rue Becquerel, 67087 Strasbourg Cedex 2, France

Abstract

In humans, wasting of lean body tissues occurs in many pathophysiological situations such as cancer, sepsis and trauma where starvation and undernutrition are also a frequent complicating factor. Many studies on protein metabolism have been performed in the whole body and skeletal muscles during acute wasting. Muscle atrophy during fasting is reflected by increased gene expression of protein breakdown components. Apart from muscles, several non-muscular tissues may also contribute to protein wasting in food-deprived situations. The interest in using long-term fasting as a model to study protein loss *in vivo* is that this state is notably marked by a successive low and high level of negative nitrogen balance. Long-term fasting is depicted by a brief period of adaptation (phase 1) which is followed by a period of large mobilization of lipids and of whole body protein sparing (phase 2) that ends with late enhanced protein utilization (phase 3). In this study, we investigated to what extent the gene expression of the main proteolytic system components found in various rat tissues was altered 1) during phase 2 and phase 3 characterizing prolonged fasting, 2) during refeeding and 3) in fasted animals treated with leptin. Fasting differentially up-regulated the expression of

[*] Corresponding author. Present address: Institut Pluridisciplinaire Hubert Curien, Département « Sciences Analytiques », Laboratoire de Spectrométrie de Masse Bio-Organique, CNRS-ULP UMR 7178, ECPM, 25 rue Becquerel, 67087 Strasbourg Cedex 2, France. Phone : (33) 3 90 24 26 81, Fax : (33) 3 90 24 27 81, Email : fbertile@chimie.u-strasbg.fr

proteolytic-related genes in a more pronounced manner in glycolytic and mixed (tibialis, gastrocnemius) than in oxidative (soleus and heart) muscles and in other tissues. As expected, MAFbx (muscle atrophy F-box) and MuRF-1 (muscle ring finger 1) were up-regulated in the catabolic phase 3 of fasting, but also to a lesser extent during phase 2, notably in tibialis and diaphragm. Upon refeeding, the restoration of mRNA levels close to fed rat values was rapid (1 day) whatever the previous duration of fasting. Recombinant leptin administration resulted in no change of proteolytic-related gene expressions. In conclusion, our study underlines the tissue- and phase-specific heterogeneity in the regulation of proteolytic-related genes in the fasting/refeeding paradigm.

Keywords: Prolonged food deprivation; rat tissues; gene expression; proteolysis.

ABBREVIATIONS

NEFA	non-esterified fatty acids
NPY	neuropeptide Y
C2 and C8	C2 and C8 20S proteasome subunits
Cat D and L	rat cathepsins D and L
CAPN3	muscle specific p94 calpain 3
MAFbx	muscle atrophy F-box E3 ubiquitin ligase
MuRF-1	muscle ring finger 1 E3 ubiquitin ligase
P2 and P3	phase 2 and phase 3 of fasting
L	leptin treatment

1. INTRODUCTION

In humans, wasting of lean body tissues occurs in many pathophysiological situations such as cancer, sepsis and trauma which are commonly associated with starvation and/or undernutrition. A better understanding of the mechanisms that underlie the loss of body proteins in such situations would help coping with metabolic disorders. Many studies on protein metabolism have been performed in the whole body and skeletal muscles during acute wasting. As it is characterized by a successive low and high level of negative nitrogen balance [1], long-term fasting then represents a good model to study in vivo protein loss. Three distinct metabolic phases have been described during prolonged fasting [1]. During the first phase (phase 1) which lasts only for a few hours in rats, glycogen is completely exhausted and fat stores are progressively used. The second phase (phase 2, P2) corresponds to the strong mobilization of fat stores whereas body proteins are efficiently spared. Finally the third phase (phase 3, P3) is characterized by an increasing protein catabolism. At this time, rats can still be refed successfully, at least after 3 days of P3. When the animal has reached a critical threshold in its body reserves utilization, a rise in locomotor activity is induced in rats and has been postulated to reflect an increasing drive for refeeding, therefore, anticipating a lethal depletion of energy stores.

Negative nitrogen balance is essentially marked by an increase in muscle protein degradation and/or by a decrease in muscle protein synthesis. Skeletal muscle is known to be the principal source of intracellular protein during food deprivation. However, proteins seem to be less mobilized in slow- than in fast-twitch muscles [2,3]. Total protein loss in individual muscles occurs also in heart and diaphragm as the fasting duration is extended [3,4]. In addition, fasting reduces the protein content of various tissues [2,4-6]. A negative nitrogen balance in non-muscular tissues would contribute to the whole-body protein wasting. Indeed, fasting promotes a decrease in the rate of protein synthesis in skeletal muscles, liver and intestine, and an increase in muscle and hepatic proteolysis [4,6].

Little is known of the mechanisms underlying protein wasting in these tissues. Although protein synthesis can be determined in vivo in tissues, the estimation of protein degradation remains problematic. An approach is to measure the mRNA levels for the various proteolytic systems. These measurements of gene expression can point to the molecular mechanisms potentially involved in protein wasting situations. There are multiple pathways for the degradation of proteins, including the lysosomal, Ca^{2+}-activated, and ATP-ubiquitin-dependent proteolytic pathways. In skeletal muscle, all three systems are functional, although the ATP-ubiquitin-proteasome pathway is believed to be the most important while in liver, the predominant proteolytic system is the lysosomal pathway. In this framework, previous studies have addressed, at the molecular level, the regulation of skeletal muscle proteolytic systems during various durations of fasting, but none of them has addressed the muscle-specific regulation of proteolytic-related genes during long-term food deprivation [7-12]. It has notably been reported that the three proteolytic systems found in skeletal muscle are selectively induced in P2 of fasting and coordinately upregulated in P3 of fasting [10]. At this later stage, we hypothesized that, apart skeletal muscles, several other tissues may also contribute to protein wasting. Although the gene expression of some proteinases and components of proteolytic processes has been shown to be stimulated during short-term fasting in rat small intestine and/or colon [6], there is a paucity of information on the proteolytic systems that operate in other tissues.

One aim of this study was therefore to examine putative phase- and/or tissue-specific heterogeneity in the levels of various mRNA encoding proteasome subunits (C2 and C8), cathepsins (Cat) D and L and muscle specific p94 calpain 3 (CAPN3) in various skeletal muscles differing in their fiber type composition, in heart, in diaphragm and in other tissues responding to energy depletion/repletion of long-term fasted/refed rats.

Muscle atrophy F-box (MAFbx) and muscle ring finger 1 (MuRF-1) are two recently identified genes encoding E3 ubiquitin ligases that are essential for mediating muscle atrophy [13]. Interestingly, MAFbx and MuRF-1 gene expressions have been shown to increase in gastrocnemius muscle under various conditions that are all associated with muscle atrophy, including short-term fasting (48h) in mice [11-14]. Another aim of this work was then to determine whether MAFbx and MuRF-1 expression depends on the phase of fasting.

Falling plasma leptin levels have been hypothesized to serve as a signal mediating the adaptation to fasting [15]. Leptin, by favouring lipid utilization, could be involved in the sequential mobilization of body energy reserves during fasting. One can hypothesize that leptin would therefore influence peripheral protein mobilization with either direct or indirect alterations of proteolytic systems at the molecular level. Therefore, our study also checked if the tissue-specific regulation of proteolytic-related mRNA expression is affected in fasted rats injected with recombinant leptin.

2. Materials and Methods

2.1. Experimental Animals

All animals (male Sprague Dawley rats from Iffa Credo, Lyon, France, weighing 245-250 g) were fed ad libitum on a chow diet consisting (by weight) of 50 % carbohydrate, 24 % protein and 5 % fat with free access to drinking water. They were maintained at an ambient temperature of 25 ± 1 °C with a 12h light/dark cycle (lights on from 08:00 A.M.). The rats weighed approximately 255-260 g when they were either sacrificed as control fed animals (n = 6) or submitted to a total fast (n = 48). On the first day of fasting, food was removed at 08:00 A.M., water remaining available ad libitum. Eight experimental groups of fasted rats (n = 6 each) were used. The animals were weighed daily. They were sacrificed by cervical dislocation at various stages of fasting. According to previous data [16], P2 and P3 of fasting can be identified from the changes in daily specific body mass loss (dm/m.dt). dm/m.dt results were subsequently validated by plasma metabolite and hormone determinations. Fasted laboratory animals can still be refed after the first three days of P3, thus showing that this is a reversible state [3,18]. In the current study, the experiment was stopped after 1–2 days of P3.

Control fed animals were compared to 1) rats in P2 and P3 of fasting, 2) rats refed for one day after P2 (P2R1) and P3 (P3R1) and 3) rats injected intraperitoneally with either PBS or rat recombinant leptin (R&D Systems GmbH, Wiesbaden, Germany, 25 μg twice a day, 08:00 A.M. and 08:00 P.M.) throughout their fast until P2 (P2PBS and P2L, respectively) and P3 (P3PBS and P3L, respectively).

Blood was sampled and plasma prepared and kept at -20°C until analysis. Soleus, gastrocnemius, tibialis anterior, heart, diaphragm, liver, stomach, duodenum, jejunum, lung, testis, kidney, adrenals and hypothalamus were rapidly collected, weighed when required, frozen in liquid nitrogen and stored at -80°C for later mRNA determinations. The carcass was frozen at -20°C until the determination of body protein content.

The protocol was approved by the local Ethics Committee and followed the guide of the Centre National de la Recherche Scientifique for the care and use of laboratory animals.

2.2. Nitrogen Determination and Plasma Assays

For body protein content analysis, the carcasses were ground under liquid nitrogen before being lyophilized and then ground again to obtain a fine powder. Homogeneous aliquots of the powder were used to determine nitrogen contents by the Kjeldahl protocol, using selenium as catalyzer. Nitrogen was converted to protein by multiplying by 6.25 [17,18]. Commercial kits were used to measure plasma urea and glucose (Sigma Diagnostics, St. Louis, MO, USA) and circulating NEFA (Wako Chemicals, Neuss, Germany).

Radioimmunoassay kits were used to determine plasma corticosterone (ICN Biomedicals, Costa Mesa, CA, USA) and insulin (Linco Research, St. Charles, MO, USA).

2.3. Northern Blot

Tissue total RNA was isolated as previously described [19]. RNA was resuspended in sterile water and subsequently quantitated by measuring the optical density at 260 nm. RNA samples (15 µg) were denaturated, subjected to 1.2 % agarose-formaldehyde gel electrophoresis, and transferred to positively charged nylon membranes (Roche Diagnostics, Mannheim, Germany) using a vacuum blotter (Qbiogene, Illkirch, France). RNA was fixed to the membrane by illuminating with UV light.

Prehybridization was done as previously described [20]. Membranes were hybridized overnight at 42°C with appropriate digoxigenin-labelled probes. Non-specific bound probe was removed by successive washes. Membranes were then incubated with an antibody against digoxigenin conjugated to alkaline phosphatase (Roche Diagnostics, Mannheim, Germany). Membranes were rinsed, reacted with a chemiluminescent alkaline phosphatase substrate (CDPstar, Roche Diagnostics, Mannheim, Germany) and exposed for 5-30 min at 37°C to X-ray film (Amersham Biosciences Europe, Freiburg, Germany). Membranes were then stripped and reused to perform detection of other mRNAs using specific probes. Signals were quantified by densitometry using Scion Image Software (http://scioncorp.com/).

Digoxigenin 5'-end-labelled oligonucleotide probes were prepared by Eurogentec (Seraing, Belgium). Specific probes were used to detect C2 and C8 proteasome subunits, Cat D, Cat L and CAPN3, and 18S rRNA as a control of loading [10]. The expression of muscle-specific ubiquitin ligases has been measured using the following probes for MuRF1: 5'-GAGCGGCTTGGCACTCAAGAGGAAGGTGGCCC-3' (GenBank accession no. NM_080903) and MAFbx: 5'-GATGTTGCCCACCAGCACGGACTTGCCGACTC-3' (GenBank accession no. AY059628). Finally, Neuropeptide Y mRNA expression has been determined using the following specific oligonucleotide probe: 5'-AGCCTGGTGGTGGCATGCATTGGTGGGACAGG-3' (GenBank accession no. M20373).

2.4. Calculations and Statistics

Results are given as means ± SEM. mRNA expression values correspond to the ratio of the specific mRNA to 18S rRNA signals and are expressed as a proportion of the control values (fed or P2, depending on the experiment) which are set to unity. Peritz' F-test [21] was used for statistical analysis. Significance was defined at the 95 % level.

3. Results

3.1. Characteristics of Fed, Refed and Fasted Rats

The initial body mass of rats (approximately 260 g) did not differ among groups. According to previous data [16], P2 and P3 of fasting were first characterized from the changes in daily loss in body mass (dm/m.dt, table 1) and later on from changes in plasma parameters (see below). dm/m.dt remained at low levels during P2 and was increased (1.4-fold) in P3. Compared to fed animals, the final body mass was decreased by nutritional treatment in P2 (1.4-fold) and in P3 (1.6-fold). The relative body mass loss was 1.4-fold higher in P3 compared to P2. The body protein content (table 1) was decreased in P2 (1.2-fold) and especially in P3 (1.8-fold) in comparison to the fed state, reflecting the late enhanced utilization of body proteins as energy fuels. Similar trends were observed when comparing P2, P2PBS and P2L and P3, P3PBS and P3L (not shown).

Refed rats gained weight in comparison with fasted animals (+10-13 % in P2R1 and P3R1 relative to P2 and P3, respectively). The body protein content of P2R1 rats was unchanged relative to P2 rats, while that of P3R1 animals was significantly increased compared to P3 (1.2-fold, table 1).

Tissue masses are also given in table 1. During the fasting period and relative to fed controls, muscle masses (soleus, gastrocnemius, tibialis anterior, heart, diaphragm) tended to progressively decrease by 4-24% (P2) and 15-36% (P3) while kidney, small intestine and liver gradually dropped by 19-55 % (P2) and 28-68% (P3). During the fasting protocol, no marked changes were observed in stomach, lung, testis, and adrenals.

Changes in urea and corticosterone levels (table 1) are closely associated with metabolic transitions and notably the increase in net proteolysis in fasting animals [12,13] and contribute to validate the metabolic status of animals. Compared to fed animals, uremia increased with fasting to a similar extent in P2, P2PBS and P2L and in P3, P3PBS and P3L by about 1.4- and 2.0-fold, respectively. Conversely, it decreased with refeeding by 19 % of the P2 value in P2R1 and by 26 % of the P3 value in P3R1. Plasma corticosterone levels followed the same variations, being increased by fasting in P2 groups (about 40-fold) and mainly in P3 groups (about 90-fold) in comparison with the fed state. Refeeding induced a drop in corticosteronemia, by 2.6-fold in P2R1 relative to P2 and by 7.7-fold in P3R1 relative to P3. Compared to fed values, plasma glucose and insulin levels were reduced in P2 and in P3 of fasting while circulating NEFA were significantly higher (1.3-fold) in P2 and lower (3-fold) in P3, all parameters being restored upon refeeding (data not shown).

To demonstrate that leptin was biologically active at a peripheral and central level, several parameters were measured as positive controls (table 2). Leptin is known to affect glucose metabolism [22] and neuropeptide gene expression involved in food intake regulation [23]. In the current study, plasma glucose levels were significantly decreased by 40% in P2L and 30% in P3L in comparison with P2PBS and P3PBS, respectively. As expected, there was a decrease in hypothalamic NPY mRNA levels in leptin-treated rats compared to animals in P2PBS and above all in P3PBS.

Table 1. Characteristics of fed, fasted and refed rats

	Fed	P2	P2R1	P3	P3R1
Duration of fasting (days)	-	4	4	5-6	5-6
Duration of refeeding (days)	-	-	1	-	1
Initial body mass, g	260.4 ± 0.1	261.0 ± 0.5	257.9 ± 0.8	258.9 ± 0.6	259.5 ± 1.0
Final body mass, g	260.4 ± 0.1[a]	190.6 ± 2.2[b]	210.6 ± 0.9[c]	162.9 ± 1.5[d]	183.4 ± 2.7[b]
Body protein content, g	53.8 ± 0.6[a]	45.7 ± 1.2[b]	43.7 ± 0.5[b]	29.4 ± 0.7[c]	39.2 ± 1.2[d]
Relative body mass loss, %	-	26.8 ± 0.8[a]	-	37.6 ± 0.6[b]	-
Relative body protein loss, %	-	13.5 ± 1.2[a]	-	45.2 ± 1.8[b]	-
dm/m.dt, g. $100g^{-1}.24h^{-1}$	-	7.13 ± 0.18[a]	-	9.80 ± 0.06[b]	-
SOL mass, mg	241 ± 9[a]	231 ± 6[a]	-	204 ± 5[b]	-
Heart mass, mg	1090 ± 31[a]	1010 ± 29[a]	-	733 ± 24[b]	-
TIB mass, mg	988 ± 55[a]	851 ± 11[a]	-	737 ± 18[b]	-
DIA mass, mg	759 ± 27[a]	577 ± 17[b]	-	483 ± 11[c]	-
Liver, g	11.90 ± 0.37[a]	5.28 ± 0.14[b]	-	3.79 ± 0.31[c]	-
Small intestine, g	6.39 ± 0.12[a]	3.63 ± 0.07[b]	-	3.02 ± 0.11[c]	-
Stomach mass, mg	1274 ± 51	1188 ± 72	-	1173 ± 50	-
Kidney, g	2.15 ± 0.08 [a]	1.74 ± 0.08 [b]		1.55 ± 0.04 [b]	
Adrenals mass, mg	57 ± 2	62 ± 2	-	57 ± 2	-
Testis mass, g	2.66 ± 0.04	2.78 ± 0.05	-	2.77 ± 0.08	-
Lung mass, g	1.38 ± 0.04	1.28 ± 0.11	-	1.17 ± 0.07	-
Corticosterone, $ng.ml^{-1}$	8.1 ± 0.9[a]	335.2 ± 18.4[b]	126.3 ± 26.9[c]	740.8 ± 55.7[d]	95.4 ± 7.8[c]
Urea, mM	8.8 ± 0.1[a]	11.7 ± 0.3[b]	9.5 ± 0.4[a]	17.1 ± 0.8[c]	12.6 ± 0.4[b]

Table 2. Plasma glucose and hypothalamic NPY mRNA levels of fasted rats injected intraperitoneally with PBS or recombinant leptin

	P2PBS	P2L	P3PBS	P3L
Glucose, mM	1.00 ± 0.03[a]	0.61 ± 0.04[b]	1.00 ± 0.08[a]	0.70 ± 0.04[b]
NPY mRNA, a.u	1.00 ± 0.13[a]	0.64 ± 0.05[b]	1.00 ± 0.17[a]	0.24 ± 0.02[b]

Results are means ± SEM of 6 determinations. Values in the same line that do not share the same superscript letter are significantly different ($P < 0.05$). P2 and P3: phase two and phase three of fasting; R1: one-day refed rats; SOL, TIB and DIA: soleus, tibialis anterior and diaphragm muscles.

Hypothalamus samples were assayed for the mRNA levels of neuropeptide Y (NPY). NPY mRNA expression data were normalized using 18S rRNA. Data were set at 1.00 in PBS animals. Results are means ± SEM of 4-6 determinations. Values in the same line and for a given stage of fasting that do not share the same superscript letter are significantly different ($P < 0.05$). P2 and P3: phase two and phase three of fasting; L: leptin-treated animals (50 µg/d).

3.2. mRNA Levels of Proteolytic-Related Genes

Northern blot analyses are illustrated in figures 1-2 and also presented in tables 3-4. As shown in table 3 and in comparison with the fed group, tibialis anterior and diaphragm exhibited a more marked increase in Cat D, C2 and C8 mRNA expressions in P3 of fasting (respectively 7-10-fold, 4.5-5-fold and 3.5-14-fold) than did soleus and/or heart (respectively 1.5-3-fold, 1.8-fold and 2.3-fold). Cat D, C2 and C8 gene expression was significantly higher in P3 compared to P2 in tibialis and diaphragm but not in soleus and heart. Moreover, CAPN3 mRNA levels were increased only in tibialis in P2 (1.4-fold) and especially in P3 (2.2-fold) compared to the fed state. Except in heart, Cat L mRNA levels were increased to a similar extent in all muscles of P2 and P3 rats (3-6-fold) in comparison with fed animals. Finally, among the genes encoding components of proteolytic systems that were studied, the gene expression of E3 ubiquitin ligases (figure 1) were the most stimulated in all muscles during fasting compared to control fed values, with MuRF-1 and MAFbx mRNA levels being significantly increased in tibialis and diaphragm (8-12-fold in P2 and 16-25-fold in P3) and to a lesser extent in soleus and heart (3-6-fold in P2 and 10-12-fold in P3). Apart from Cat L, proteolytic-related gene expressions seem then to be more affected by fasting in fast-twitch skeletal muscle and diaphragm than in slow-twitch skeletal muscle and heart. During refeeding, as shown in figure 1 for MAFbx and MuRF-1, mRNA levels were rapidly decreased in all muscles in comparison with fasting. Most of the values measured in the P2R1 and P3R1 groups were similar to those of the fed group. For C2, C8, Cat L, Cat D and CAPN3, similar results were obtained upon refeeding (data not shown).

mRNA levels of C2 and Cat L in liver, duodenum and jejunum, kidney, stomach, adrenals, testis, and lung are presented in table 4. C2 mRNA levels were unchanged in both P2 and P3 in comparison with the fed group, except for a slight increase (1.4- to 1.5-fold in P3) in testis and liver. In contrast, C2 gene expression rose in kidney (1.9-fold in P2 and P3). For a given tissue, C3 and Cat D gene expressions (data not shown) were roughly similar to those of C2 during the long-term fasting protocol.

During fasting and compared to fed values, Cat L mRNA levels were 1) unaffected in stomach and adrenals, 2) slightly increased in liver in P2 and P3 (1.5- to 1.7-fold) and in testis, lung and kidney only in P3 (1.5- to 1.6-fold). In contrast, Cat L gene expression rose in duodenum and jejunum in P2 (2.2- and 1.6-fold, respectively) and especially in P3 (4.8- and 3.0-fold, respectively) compared to fed values.

Table 3. mRNA levels of proteolytic-related genes in muscles of fed and fasted rats.

		Soleus	Heart	Tibialis	Diaphragm
C2	Fed	1.00 ± 0.10^{a}	1.00 ± 0.09^{a}	1.00 ± 0.07^{a}	1.00 ± 0.12^{a}
	P2	1.76 ± 0.07^{b}	1.39 ± 0.05^{b}	1.88 ± 0.14^{b}	1.73 ± 0.26^{b}
	P3	1.80 ± 0.17^{b}	$1.23 \pm 0.05^{a,b}$	5.03 ± 0.60^{c}	4.49 ± 0.48^{c}
C8	Fed	1.00 ± 0.13	1.00 ± 0.07^{a}	1.00 ± 0.12^{a}	1.00 ± 0.11^{a}
		Soleus	Heart	Tibialis	Diaphragm
	P2	1.61 ± 0.32	1.98 ± 0.20^{b}	1.61 ± 0.18^{b}	3.61 ± 0.46^{b}
	P3	1.72 ± 0.32	2.29 ± 0.55^{b}	3.56 ± 0.52^{c}	13.68 ± 2.75^{c}
Cat D	Fed	1.00 ± 0.07^{a}	1.00 ± 0.06^{a}	1.00 ± 0.08^{a}	1.00 ± 0.05^{a}
	P2	1.74 ± 0.23^{b}	1.40 ± 0.05^{b}	3.43 ± 0.44^{b}	4.50 ± 0.86^{b}
	P3	2.91 ± 0.61^{b}	1.46 ± 0.13^{b}	10.18 ± 1.60^{c}	7.41 ± 0.91^{c}
Cat L	Fed	1.00 ± 0.10^{a}	1.00 ± 0.03^{a}	1.00 ± 0.11^{a}	1.00 ± 0.11^{a}
	P2	3.12 ± 0.48^{b}	2.83 ± 0.19^{b}	2.72 ± 0.34^{b}	2.41 ± 0.32^{b}
	P3	5.89 ± 0.77^{c}	3.25 ± 0.19^{b}	5.86 ± 0.96^{c}	5.01 ± 0.54^{c}
CAPN3	Fed	1.00 ± 0.09	-	1.00 ± 0.13^{a}	1.00 ± 0.07
	P2	1.35 ± 0.15	-	1.46 ± 0.09^{b}	0.91 ± 0.06
	P3	1.50 ± 0.20	-	2.24 ± 0.11^{c}	1.05 ± 0.28

Muscle samples were assayed for the mRNA levels of C2 and C8 proteasome subunits, cathepsins (Cat) D and L and muscle specific p94 calpain 3 (CAPN3). Data were normalized using 18S rRNA and the abundance was set at 1.00 in fed animals. Results are means ± SEM of 5-6 determinations. Within a same column for a given proteolytic-related component, values that do not share the same superscript letter are significantly different ($P < 0.05$). P2 and P3: phase two and phase three of fasting.

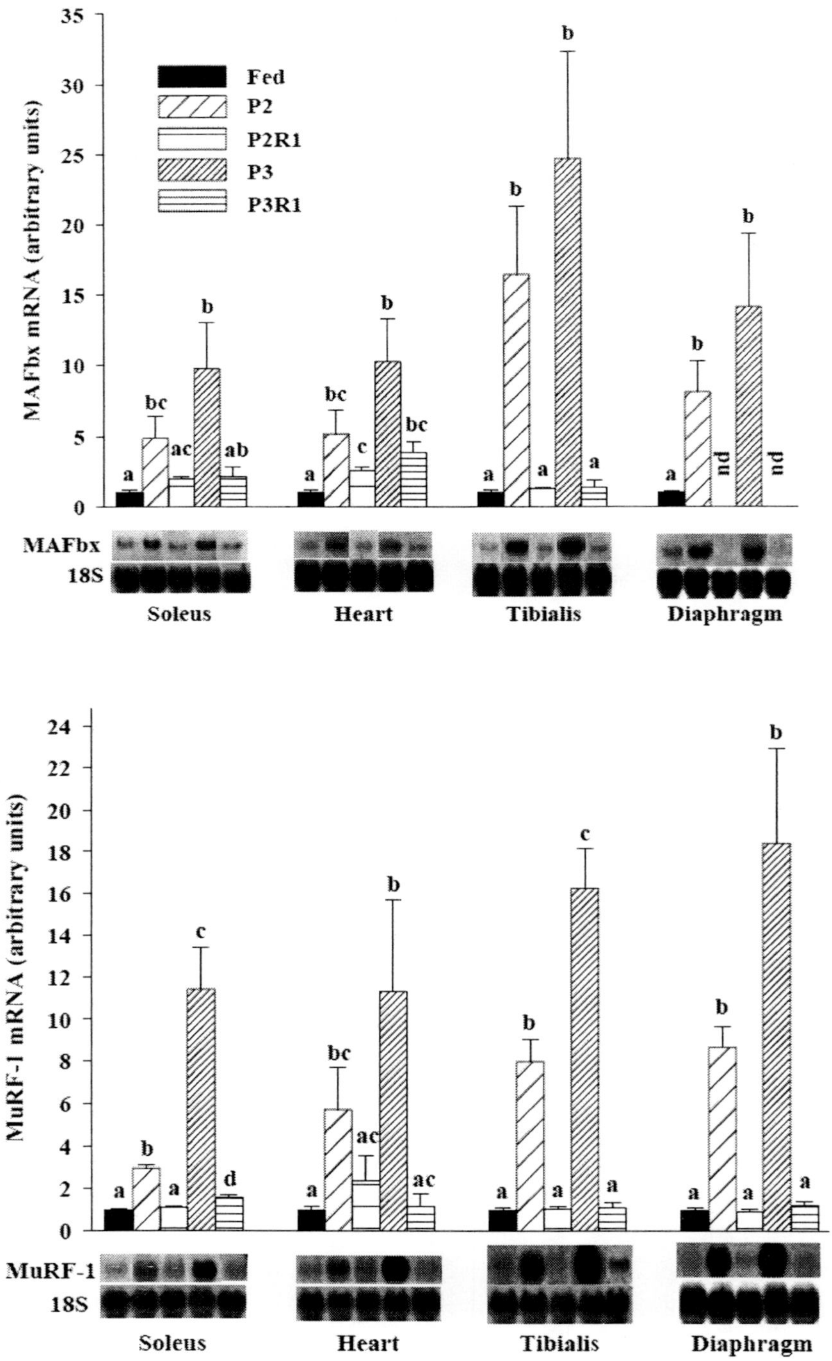

Figure 1. MAFbx and MuRF-1 mRNA levels in soleus, heart, tibialis and diaphragm muscles of fed, fasted and refed rats. Fasted rats were studied in phase 2 (P2) and phase 3 (P3) and refed rats received food *ad libitum* for one day after P2 (P2R1) or P3 (P3R1). mRNA levels were determined by Northern blot analysis. Data were normalized using 18S rRNA and the abundance was set at 1.0 for fed animals. Results are means ± SEM of 5-6 determinations. Representative Northern blots are shown below the corresponding bars. Values that do not share the same superscript letter are significantly different ($P < 0.05$). MAFbx and MuRF-1, muscle atrophy F-box and muscle ring finger 1 E3 ubiquitin ligases.

Figure 2 depicts the gene expression of C2 and Cat L in gastrocnemius muscle that has been mentioned to be among the most recruited during fasting. As expected, fasting induced an up-regulation of C2 (4.3-fold in P3) and Cat L (3.8- and 18.6-fold in P2 and P3, respectively) mRNA levels in gastrocnemius in comparison with the fed state. Moreover, no difference in C2 and Cat L mRNA levels was observed between P2PBS and P2L or between P3PBS and P3L in gastrocnemius (figure 2). Similar trends have been obtained for other tissues (data not shown). Likewise, leptin treatment did not alter Cat D gene expressions in any of the tissues considered (data not shown).

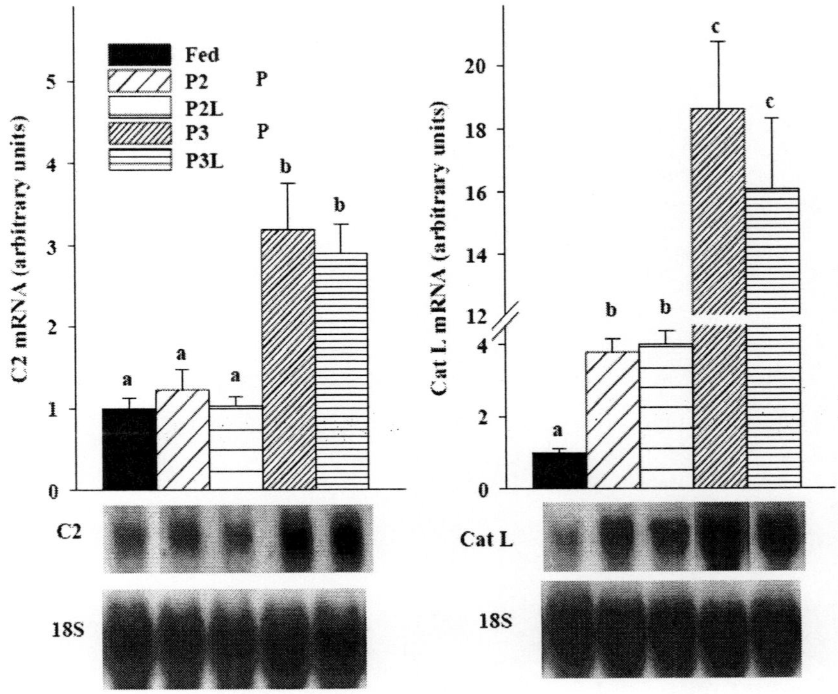

Figure 2. C2 and Cat L mRNA levels in gastrocnemius muscle of fed and fasted rats injected intraperitoneally with either PBS or 50µg/d of recombinant leptin (L). Fasted rats were studied in phase 2 (P2) and phase 3 (P3) of fasting. mRNA levels were determined by Northern blot analysis. Data were normalized using 18S rRNA and the abundance was set at 1.0 for fed animals. Results are means ± SEM of 5-6 determinations. Representative Northern blots are shown below the corresponding bars. Values that do not share the same superscript letter are significantly different ($P < 0.05$). C2, 20S proteasome subunit; Cat L, cathepsin L.

Table 4. mRNA levels of proteolytic-related genes in various tissues of fed and fasted rats

		Liver	Duodenum	Jejunum	Kidney	Stomach	Adrenals	Testis	Lung
C2	Fed	1.00 ± 0.06^{a}	1.00 ± 0.10	1.00 ± 0.05	1.00 ± 0.08^{a}	1.00 ± 0.05	1.00 ± 0.06	1.00 ± 0.07^{a}	1.00 ± 0.03
	P2	1.16 ± 0.06^{a}	1.10 ± 0.06	0.94 ± 0.13	1.91 ± 0.25^{b}	0.96 ± 0.08	1.08 ± 0.06	1.01 ± 0.08^{a}	1.14 ± 0.05
	P3	1.51 ± 0.11^{b}	1.16 ± 0.14	1.07 ± 0.12	1.90 ± 0.22^{b}	0.78 ± 0.09	1.06 ± 0.04	1.39 ± 0.09^{b}	1.15 ± 0.07
Cat L	Fed	1.00 ± 0.03^{a}	1.00 ± 0.09^{a}	1.00 ± 0.11^{a}	1.00 ± 0.06^{a}	1.00 ± 0.06	1.00 ± 0.04	1.00 ± 0.04^{a}	1.00 ± 0.09^{a}
	P2	1.52 ± 0.10^{b}	2.23 ± 0.26^{b}	1.64 ± 0.15^{b}	$1.15 \pm 0.05^{a,b}$	1.04 ± 0.10	1.04 ± 0.09	1.13 ± 0.09^{a}	1.18 ± 0.04^{a}
	P3	1.74 ± 0.12^{b}	4.85 ± 0.57^{c}	2.99 ± 0.29^{c}	1.58 ± 0.19^{b}	1.27 ± 0.09	1.12 ± 0.03	1.54 ± 0.13^{b}	1.59 ± 0.14^{b}

Tissue samples were assayed for the mRNA levels of C2 proteasome subunit and cathepsin (Cat) L. Data were normalized using 18S rRNA and the abundance was set at 1.00 in fed animals. Results are means ± SEM of 5-6 determinations. Within a same column for a given proteolytic-related component, values that do not share the same superscript letter are significantly different ($P < 0.05$). P2 and P3: phase two and phase three of fasting.

4. Discussion

The mechanisms responsible for protein wasting in various tissues is mediated by decreased protein synthesis and/or increased protein breakdown. Data dealing with whole-body and tissue protein synthesis has already been reported during brief and prolonged fasting in the rat [4]. The current study examines putative tissue-specific heterogeneity in the expression of genes encoding components of proteolytic systems after nutritional manipulations and leptin treatment in various tissues of long-term fasted/refed rats. The measurement of mRNA expression for various components of proteolytic systems constitutes an approach that has been extensively used for skeletal muscle in order to estimate protein degradation. In rodent muscle, these mRNA levels highly correlate with rates of protein degradation measured in vitro [8,24-29], with proteolytic protein levels detected by western blotting [30] and with proteinase activities [26,31]. Furthermore, increased expression of proteolytic system components are observed in the muscles from cachectic patients and correlate with negative nitrogen balance [32,33] and with increased urinary 3-methylhistidine excretion, which is an index of enhanced myofibrillar proteolysis [33].

During fasting, muscle protein degradation is essential to provide the organism with amino acids for gluconeogenesis [34]. Early reports have shown that rat muscles containing predominantly slow-twitch oxidative fibers (soleus) lose little mass and protein during fasting (relative sparing) when compared to muscles composed mainly of either fast-twitch oxidative glycolytic or fast-twitch glycolytic fibers (extensor digitorum longus, epitrochlearis) [2,3]. It has also been reported that the magnitude of total protein loss during long-term fasting in heart is similar to that in extensor digitorum longus, a fast oxidative glycolytic muscle [3]. We show here that long-term fasting in rats resulted in up-regulated muscle expression of critical components of the three proteolytic systems in P2 and especially in P3, these changes affecting mainly a specific group of muscles including fast glycolytic gastrocnemius, fast oxidative-glycolytic tibialis anterior and diaphragm to a lesser extent. In contrast, fasted rats exhibited much less or even no stimulation of these gene expressions in slow oxidative soleus and in heart. Thus, the coordinate up-regulation of all studied proteolytic-related genes that we reported previously in gastrocnemius [10] is not an invariable response found in all muscles. The differences among muscles may reflect an effect of composing fiber types and/or differences in energy metabolism, with glycolytic and/or mixed muscles being likely to be more sensitive to fasting than oxidative muscles (soleus and heart). This is consistent with the much more pronounced atrophy of mixed-fiber than that of slow-twitch muscles in various other catabolic situations, including sepsis [31], cancer [35], diabetes [36] and glucocorticoid administration [37].

MAFbx and MuRF-1 are two genes encoding E3 ubiquitin ligases that were recently described to be expressed only in skeletal muscles and up-regulated in various models of atrophy including short-term fasting [12-14,38]. Previous results have shown that the experimental over-expression of MAFbx in myotubes produces atrophy [13]. MAFbx and MuRF-1 up-regulation would then constitute essential steps prior to muscle atrophy and protein degradation. In addition, MAFbx and MuRF-1 deficiencies in mice trigger significant sparing of muscle mass when muscle atrophy is experimentally induced [13]. Therefore, the selective up-regulation of these genes could be at the origin of the specific muscle atrophy during long-term food deprivation. In the current study, MAFbx and MuRF-1 mRNAs were more strongly up-regulated by fasting than mRNAs encoding other components of the three main proteolytic pathways. Moreover, high levels of MAFbx and MuRF-1 expression were found, in P2 and even more in P3 of fasting, in all muscles in the following order: tibialis and diaphragm > heart > soleus. Overall, these results support the view that 1) MAFbx and MuRF-1 would represent sensitive markers of muscle protein loss in these catabolic situations, and 2) tissue-specific MAFbx and MuRF-1 expression is in line with the differential protein wasting in muscles, oxidative skeletal muscles (soleus) losing less protein than other muscles during fasting [2,3].

Little is known regarding the tissue-specific expression of proteolytic-related genes during long-term food deprivation. This is a major gap because crucial functions have to be maintained in stressful situations such as prolonged fasting to promote animal survival. Liver plays a crucial role in energy metabolism. Apart from the kidney's roles in excreting waste material from the bloodstream, secreting hormones and maintaining serum electrolytes, acid-base levels and osmolality, it has been shown that renal glucose production may account for 20-25 % of whole-body glucose turnover after 60h of fasting in humans [39]. Thus, the maintenance of hepatic and renal functions would be of great importance because gluconeogenesis notably constitutes an essential mechanism for metabolic adjustments during fasting. Together with the stomach, the gut is a central organ in overall metabolism [40]. The small intestine has one of the highest rates of protein turnover of any tissue in the body [40]. Small intestine gluconeogenesis contributes for 35 % of the total glucose production in 72h-fasted rats [41]. It has already been reported that mRNA levels for most components of the lysosomal, Ca^{2+}-activated and ubiquitin-dependent proteolytic pathways are increased in whole small intestine of adult rats in P2 of fasting [6]. However, lacking data on putative differences between 1) duodenum and jejunum, and 2) P2 and P3 of fasting would be interesting because intestinal protein wasting during long-term food deprivation may promote gut barrier dysfunction and compromise digestive and absorptive functions when refeeding. Moreover, in stressful situations, such as prolonged fasting, the adrenocortical responsiveness has to be maintained and adrenals are then expected to be relatively spared. A reverse trend is expected for testis since the hypothalamic-pituitary-gonadal axis is known to be inhibited during fasting [42]. In the current study, we show that fasting-induced elevated mRNA levels of proteolytic-related genes is much less marked in non-muscular tissues than in muscles. Moreover, gene expressions in tissues other than muscles were either only slightly augmented or similar in P3 relative to P2 of fasting. This suggests that even if protein breakdown of non-muscular tissues may participate in providing amino acids for energy needs during fasting, it is unlikely to be involved to the extent of the muscular protein catabolism in the late enhanced proteolysis.

Enhanced protein breakdown, reflected by elevated mRNA levels of components of the ubiquitin- and Ca^{2+}-dependent proteolytic systems, has been suggested to be necessary in the early recovery from muscle atrophy in reloaded unweighted soleus muscle [29]. On the contrary, it has been shown recently in skeletal muscle that recovery from fasting (48h) involves the restoration to control fed values of the expression of C2 and C9 proteasome subunits after 24h of refeeding [43]. In the current study, we extend these data by showing that the rapid restoration (24h) of the levels of proteolytic-related mRNA levels close to fed rat values occurred for components of the three main proteolytic pathways in skeletal muscles, thus promoting efficient protein replenishment. However, the body protein content of P2R1 rats was unchanged relative to P2 rats, while that of P3R1 animals was significantly increased compared to P3 (table 1). Moreover, the proportion of protein expressed relative to the fresh mass (g protein/100g) was decreased in P2R1 rats (20.5 ± 0.3 %) compared to P2 (23.8 ± 0.8 %) animals and augmented in P3R1 (20.9 ± 0.8 %) in comparison to P3 (17.4 ± 0.4 %). These data suggest that the lipid and carbohydrate compartments might be restored preferentially after P2 and that restoration of the protein compartment seems to occur earlier after P3 than after P2 by mechanisms possibly involving selective rates of protein synthesis according to the previous duration of fasting.

Skeletal muscle carbohydrate and fat metabolisms are altered by leptin [15]. Previous data have shown that 1) leptin does not seem to have a major role in the lean body mass loss observed in stressed rats fed a high-fat diet [44] and 2) leptin treatment in fed rodents neither significantly affected protein degradation and synthesis nor altered the expression of several genes involved in muscle proteolysis, as compared with the control group [45,46]. We show here that recombinant leptin, administered in order to prevent the fasting-induced decreased plasma levels of the protein, resulted in no change of proteolytic-related gene expressions during long-term fasting (figure 2).

5. CONCLUSION

In conclusion, our study by reporting the tissue- and phase-specific heterogeneity in the regulation of proteolytic-related genes provides further insights into the molecular mechanisms involved in protein wasting in the prolonged fasting paradigm. Fasting affects especially fast glycolytic and fast-oxidative glycolytic skeletal muscles and to a lesser extent slow-oxidative muscles while other tissues would be relatively spared. This would allow for protection of crucial functions. During fasting, MAFbx and MuRF-1 up-regulation would constitute a necessary and essential step prior to muscle protein degradation. As expected, MAFbx and MuRF-1 are markedly up-regulated in muscle in the catabolic P3 of fasting, but also during P2 which has been described as a phase of low protein utilization indicating that these markers are very sensitive to nutritional changes. The rapid restoration of mRNA levels close to fed rat values occurs upon one day of refeeding. To what extent this may promote selective protein replenishment according to the previous depletion and/or relative loss of body protein and fat would need close examinations to better understand the restoration of the whole body condition. Further time-course studies will have to check whether the protein wasting in some tissues such as the gastrointestinal tract resulted from decreased protein synthesis, cell loss into the gut lumen or enhanced secretion of proteins such as digestives enzymes and apolipoproteins. Among the various changes that occur between P2 and P3 of fasting, the identification and role of signals other than glucocorticoids that trigger protein degradation when reaching P3 should also be clarified.

ACKNOWLEDGMENTS

We would like to thank A. Wulgué for her skilled technical assistance. This study was supported by grants from the Bettencourt Schueller Fund (to F.B.) and from the Centre National de la Recherche Scientifique (to T.R.).

REFERENCES

[1] Goodman, MN; Larsen, PR; Kaplan, MM; Aoki, TT; Young, VR; Ruderman, NB. Starvation in the rat. II. Effect of age and obesity on protein sparing and fuel metabolism. *Am. J. Physiol,* 1980 239, E277-E286.

[2] Goodman, MN; Lowell, B; Belur, E; Ruderman, NB. Sites of protein conservation and loss during starvation: influence of adiposity. *Am. J. Physiol,* 1984 246, E383-90.

[3] Cherel, Y; Robin, JP; Heitz, A; Calgari, C; Le Maho, Y. Relationships between lipid availability and protein utilization during prolonged fasting. *J. Comp. Physiol. B,* 1992 162, 305-313.

[4] Cherel, Y; Attaix, D; Rosolowska-Huszcz, D; Belkhou, R; Robin, JP; Arnal, M; Le Maho, Y. Whole-body and tissue protein synthesis during brief and prolonged fasting in the rat. *Clin. Sci. (Lond)* , 1991 81, 611-9.

[5] Cherel, Y; Le Maho, Y. Refeeding after the late increase in nitrogen excretion during prolonged fasting in the rat. *Physiol. Behav,* 1991 50, 345-9.

[6] Samuels, SE; Taillandier, D; Aurousseau, E; Cherel, Y; Le Maho, Y; Arnal, M; Attaix, D. Gastrointestinal tract protein synthesis and mRNA levels for proteolytic systems in adult fasted rats. *Am. J. Physiol,* 1996 271, E232-E238.

[7] Jagoe, RT; Lecker, SH; Gomes, M; Goldberg, AL. Patterns of gene expression in atrophying skeletal muscles: response to food deprivation. *FASEB J,* 2002 16, 1697-1712.

[8] Medina, R; Wing, SS; Goldberg, AL. Increase in levels of polyubiquitin and proteasome mRNA in skeletal muscle during starvation and denervation atrophy. *Biochem. J,* 1995 307 (Pt 3), 631-7.

[9] Ilian, MA; Forsberg, NE. Gene expression of calpains and their specific endogenous inhibitor, calpastatin, in skeletal muscle of fed and fasted rabbits. *Biochem. J,* 1992 287 (Pt 1), 163-71.

[10] Bertile, F; Le Maho Y.; Raclot, T. Coordinate upregulation of proteolytic-related genes in rat muscle during late fasting. *Biochem. Biophys. Res. Commun,* 2003 311, 929-934.

[11] Lecker, SH; Jagoe, RT; Gilbert, A; Gomes, M; Baracos, V; Bailey, J; Price, SR; Mitch W.E.; Goldberg, AL. Multiple types of skeletal muscle atrophy involve a common program of changes in gene expression . *FASEB J,* 2004 18, 39-51.

[12] Gomes, MD; Lecker, SH; Jagoe, RT; Navon, A; Goldberg, AL. Atrogin-1, a muscle-specific F-box protein highly expressed during muscle atrophy. *Proc. Natl. Acad. Sci. U S A,* 2001 98, 14440-14445.

[13] Bodine, SC; Latres, E; Baumhueter, S; Lai, VK; Nunez, L; Clarke, BA; Poueymirou, WT; Panaro, FJ; Na, E; Dharmarajan, K; Pan, ZQ; Valenzuela, DM; DeChiara, TM;

Stitt, TN; Yancopoulos, GD; Glass, DJ. Identification of ubiquitin ligases required for skeletal muscle atrophy. *Science,* 2001 294, 1704-1708.

[14] Wray, CJ; Mammen, JMV; Hershko, DD; Hasselgren, P-O. Sepsis upregulates the gene expression of multiple ubiquitin ligases in skeletal muscle. *Int. J. Biochem. Cell Biol,* 2003 35, 698-705.

[15] Ahima, RS; Flier, JS. Leptin. *Annu. Rev. Physiol,* 2000 62, 413-37.

[16] Cherel, Y; Robin, JP; Le Maho, Y. Physiology and biochemistry of long term fasting in birds. *Can. J. Zool,* 1988 66, 159-66.

[17] Challet, E; Le Maho, Y; Robin, JP; Malan, A; Cherel, Y. Involvement of corticosterone in the fasting-induced rise in protein utilization and locomotor activity. *Pharmacol. Biochem. Behav*, 1995 50, 405-12.

[18] Belkhou, R; Cherel, Y; Heitz, A; Robin, JP; Le Maho, Y. Energy contribution of proteins and lipids during prolonged fasting in the rat. *Nutrition research,* 1991 11, 365-74.

[19] Chomczynski, P; Sacchi, N. Single-step method of RNA isolation by acid guanidinium thiocyanate- phenol-chloroform extraction. *Anal. Biochem,* 1987 162, 156-9.

[20] Trayhurn, P; Duncan, JS; Nestor, A; Thomas, ME; Eastmond, NC; Rayner, DV. Rapid chemiluminescent detection of mRNAs on northern blots with digoxigenin end-labelled oligonucleotides. *Electrophoresis,* 1995 16, 341-4.

[21] Harper, JF. Peritz' F test: basic program of a robust multiple comparison test for statistical analysis of all differences among group means. *Comput. Biol. Med,* 1984 14, 437-45.

[22] Kamohara, S; Burcelin, R; Halaas, JL; Friedman, JM; Charron, MJ. Acute stimulation of glucose metabolism in mice by leptin treatment. *Nature,* 1997 389, 374-7.

[23] Schwartz, MW; Seeley, RJ; Campfield, LA; Burn, P; Baskin, DG. Identification of targets of leptin action in rat hypothalamus. *J. Clin. Invest,* 1996 98, 1101-6.

[24] Voisin, L; Breuille, D; Combaret, L; Pouyet, C; Taillandier, D; Aurousseau, E; Obled, C; Attaix, D. Muscle wasting in a rat model of long-lasting sepsis results from the activation of lysosomal, Ca2+ -activated, and ubiquitin-proteasome proteolytic pathways. *J. Clin. Invest,* 1996 97, 1610-7.

[25] Wing, SS; Goldberg, AL. Glucocorticoids activate the ATP-ubiquitin-dependent proteolytic system in skeletal muscle during fasting. *Am. J. Physiol,* 1993 264, E668-76.

[26] Taillandier, D; Aurousseau, E; Meynial-Denis, D; Bechet, D; Ferrara, M; Cottin, P; Ducastaing, A; Bigard, X; Guezennec, CY; Schmid, HP; Attaix, D. Coordinate activation of lysosomal, Ca 2+-activated and ATP-ubiquitin- dependent proteinases in the unweighted rat soleus muscle. *Biochem. J,* 1996 316 (Pt 1), 65-72.

[27] Temparis, S; Asensi, M; Taillandier, D; Aurousseau, E; Larbaud, D; Obled, A; Bechet, D; Ferrara, M; Estrela, JM; Attaix, D. Increased ATP-ubiquitin-dependent proteolysis in skeletal muscles of tumor-bearing rats. *Cancer Res,* 1994 54, 5568-73.

[28] Price, SR; England, BK; Bailey, JL; Van Vreede, K; Mitch, WE. Acidosis and glucocorticoids concomitantly increase ubiquitin and proteasome subunit mRNAs in rat muscle. *Am. J. Physiol,* 1994 267, C955-60.

[29] Taillandier, D; Aurousseau, E; Combaret, L; Guezennec, C-Y; Attaix, D. Regulation of proteolysis during reloading of the unweighted soleus muscle. *Int. J. Biochem. Cell Biol,* 2003 35, 665-675.

[30] Khal, J; Hine, AV; Fearon, KCH; Dejong, CHC; Tisdale, MJ. Increased expression of proteasome subunits in skeletal muscle of cancer patients with weight loss. *Int. J. Biochem. Cell Biol,* 2005 37, 2196-2206.

[31] Hobler, SC; Williams, A; Fischer, D; Wang, JJ; Sun, X; Fischer, JE; Monaco, JJ; Hasselgren, PO. Activity and expression of the 20S proteasome are increased in skeletal muscle during sepsis. *Am. J. Physiol,* 1999 277, R434-R440.

[32] Tiao, G; Hobler, S; Wang, JJ; Meyer, TA; Luchette, FA; Fischer, JE; Hasselgren, PO. Sepsis is associated with increased mRNAs of the ubiquitin-proteasome proteolytic pathway in human skeletal muscle. *J. Clin. Invest,* 1997 99, 163-168.

[33] Mansoor, O; Beaufrere, B; Boirie, Y; Ralliere, C; Taillandier, D; Aurousseau, E; Schoeffler, P; Arnal, M; Attaix D. Increased mRNA levels for components of the lysosomal, Ca2+-activated, and ATP-ubiquitin-dependent proteolytic pathways in skeletal muscle from head trauma patients. *Proc. Natl. Acad. Sci. U S A,* 1996 93, 2714-2718.

[34] Kettelhut, IC; Wing, SS; Goldberg, AL. Endocrine regulation of protein breakdown in skeletal muscle. *Diabetes Metab. Rev,* 1988 4, 751-772.

[35] Baracos, VE; DeVivo, C; Hoyle, DH; Goldberg, AL. Activation of the ATP-ubiquitin-proteasome pathway in skeletal muscle of cachectic rats bearing a hepatoma. *Am. J. Physiol,* 1995 268, E996-1006.

[36] Pepato, MT; Migliorini, RH; Goldberg, AL; Kettelhut, IC. Role of different proteolytic pathways in degradation of muscle protein from streptozotocin-diabetic rats. *Am. J. Physiol,* 1996 271, E340-E347.

[37] Savary, I; Debras, E; Dardevet, D; Sornet, C; Capitan, P; Prugnaud, J; Mirand, PP; Grizard, J. Effect of glucocorticoid excess on skeletal muscle and heart protein synthesis in adult and old rats. *Br. J. Nutr,* 1998 79, 197-304.

[38] Wing, SS. Control of ubiquitination in skeletal muscle wasting. *Int. J. Biochem. Cell Biol,* 2005 37, 2075-2087.

[39] Ekberg, K; Landau, BR; Wajngot, A; Chandramouli, V; Efendic, S; Brunengraber, H; Wahren, J. Contributions by kidney and liver to glucose production in the postabsorptive state and after 60 h of fasting. *Diabetes,* 1999 48, 292-298.

[40] McBurney, MI. The gut: central organ in nutrient requirements and metabolism. *Can. J. Physiol. Pharmacol,* 1994 72, 260-265.

[41] Mithieux, G; Bady, I; Gautier, A; Croset, M; Rajas, F; Zitoun, C. Induction of control genes in intestinal gluconeogenesis is sequential during fasting and maximal in diabetes. *Am. J. Physiol. Endocrinol. Metab,* 2004 286, E370-E375.

[42] Ahima, RS; Prabakaran, D; Mantzoros, C; Qu, D; Lowell, B; Maratos-Flier, E; Flier, JS. Role of leptin in the neuroendocrine response to fasting. *Nature,* 1996 382, 250-2.

[43] Kee, AJ; Combaret, L; Tilignac, T; Souweine, B; Aurousseau, E; Dalle, M; Taillandier, x; Attaix, D. Ubiquitin-proteasome-dependent muscle proteolysis responds slowly to insulin release and refeeding in starved rats. *J. Physiol. - London,* 2003 546, 765-776.

[44] Harris, RB; Zhou, J; Youngblood, BD; Rybkin, II; Smagin, GN; Ryan, DH. Effect of repeated stress on body weight and body composition of rats fed low- and high-fat diets. *Am. J. Physiol,* 1998 275, R1928-R1938.

[45] Kaibara, A; Moshyedi, A; Auffenberg, T; Abouhamze, A; Copeland, EM 3rd; Kalra, S; Moldawer, LL. Leptin produces anorexia and weight loss without inducing an acute phase response or protein wasting. *Am. J. Physiol,* 1998 274, R1518-25.

[46] Carbo, N; Ribas, VV; Busquets, S; Alvarez, B; Lopez-Soriano, FJ; Argiles, JM. Short-term effects of leptin on skeletal muscle protein metabolism in the rat. *J. Nutr. Biochem,* 2000 11, 431-435.

In: Messenger RNA Research Perspectives
Editor: Takeo Takeyama, pp. 115-123

ISBN: 978-1-60021-814-9

Chapter IV

Gene Markers for Renal Cell Carcinoma and their Clinical Applications

Guorong Li*[*]*, Gang Feng, Christian Genin and Jacques Tostain
Department of Urology, North Hospital, CHU of Saint-Etienne, France
Clinical Immunology Laboratory, North Hospital, CHU of Saint-Etienne, France

Abstract

One major difficulty in the diagnosis and treatment of renal cell carcinoma (RCC) is the limited sensitivity of techniques for the initial detection. RCC lacks efficient serum markers. The diagnosis depends solely on the imaging analysis. Recently, many studies have focused on the discoveries of molecular markers for tumors. Several markers of messenger RNAs (gene markers) have been available for RCC. The use of messenger RNAs to solve the clinical problems in RCC is relatively new and promising. In this chapter, we describe our personal experience on the study of the messenger RNAs in renal tumors and how we used these markers in clinical applications. We used the traditional method as well as developed a convenient assay on the messenger RNA of renal tumor samples. The messenger RNAs that we have studied are: MN/CA9, cadherin-6, vimentin, S100A1, etc. We further used the gene marker in cytoponction for the differential diagnosis, in the blood for detection of circulating renal cancer cells as well as in the biopsy for subtyping of renal tumors. All these studies demonstrate that the messenger RNAs are useful means to solve the clinical problems in the field of RCC.

Keywords: gene markers, RT-PCR, renal cell carcinoma.

[*] Correspondence: Guorong Li, MD, PhD, Associate professor, North Hospital, CHU of Saint Etienne, 42055 Saint-Etienne Cedex 2, France. Phone: 0033 (0) 4 77 82 88 14; Fax: 0033 (0) 4 77 12 05 52; E-mail: grli2001@yahoo.fr

INTRODUCTION

The incidence of RCC has increased steadily during recent years. The radiological image has revolutionized the diagnosis of RCC. Many small asymptomic renal tumors have been diagnosed. Our clinical data indicated that about 20% of these small tumors are benign [1]. At present, nearly one-third of patients present metastatic disease at the time of diagnosis. Besides, about 30 - 40% of patients with local tumors will ultimately relapse with metastatic disease after surgery [2]. These clinical facts indicate that search for new diagnostic and monitoring methods are needed.

Gene markers are useful for diagnosis or prognosis of cancer. However, the clinical use of gene expression in renal cancer is limited. One of the major reasons is that the available makers for RCC are scarce. In this chapter, we use our personal experience to discuss the advances on the gene markers available for RCC and to describe their clinical applications.

TECHNICAL POINTS FOR DETECTION OF MRNA MARKERS IN RENAL TUMOR SAMPLES

The human genome contains 25000-35000 genes. Only about 15% of these genes are expressed in a given cell. Different tissues and cells express different gene subsets. Gene expression is strictly regulated under normal physiologic conditions but is often disrupted during malignant transformation and tumor progression.

The RT-PCR is an often used technique for the study of gene expression. The experimental steps include RNA extraction, reverse-transscription polymerase chain reaction (RT-PCR) and gel electrophoresis. The technique is simple and routinely available in a hospital laboratory. There are many methods to separate the RNA and different laboratories use different protocols. This certainly affects the sensitivity and specificity.

We have developed a rapid and sensitive detection of mRNA expressions for molecular diagnosis of conventional RCC [3]. Our process contains a column extraction of RNA and one-step RT-PCR. The column extraction of RNA was more sensitive than commonly used chroloform-ethanol method. Our experience showed that the quality of extracted RNA was also comparable or superior to that of chroloform-ethanol method. It needs only twenty minutes. In fact, this column extraction of RNA was recently used to extract RNA from microdissection specimens where the number of cells is limited [4, 5]. In our study, we also found that the extraction can be effective even if 10 cells were used (figure 1). However, the extraction of RNA can be much more stable at a higher number of cells. One-step RT-PCR limits the chance of possible contamination and makes the process easily standardized when a certain quantity of RNA is used for analysis. Besides, one-step RT-PCR provides high throughput RNA for amplification. The combination of column extraction of RNA and one-step RT-PCR significantly reduces labor and time consumption. The whole process can be finished within several hours.

RECENT MRNA MARKERS FOR RENAL CELL CARCINOMA

Renal cell carcinoma (RCC) is a common tumour. The common subtypes are conventional or clear cell carcinoma, papillary carcinoma, chromophobe carcinoma and benign oncocytoma [6]. Each subtype may have the different clinical evolution [7]. The exact differentiation of each subtype is of clinical importance.

For the detection of renal cancer cells, the first question is to choose a marker. Renal cancer cells express traditional markers for epithelial cells [8]. The epithelial membrane antigens present in several forms : membrane, secret and a glycoform produced by tumor cells (CA15.3). The epithelial specific antigens (ESA, Ep-CAM, HEA125 and Ber-EP4) and epithelial associated antigen MOC31 can be expressed by renal cells. Renal cancer cells can express these antigens with a lesser degree because of dedifferentiation. In the same way, renal epithelial cells can express cytoplasmic protein as found in renal cancer cells. The renal tubular cells have a grand enzymatic activity. Amino-peptidase A (gp 160), CD13 and CD 26 are expressed in the proximal tubular cells. Their expression can be preserved in renal cancer cells. Renal cancer cells can also express PSA or PSMA.

As new techniques such as the microarray and gene chip are advanced in cancer research, new gene markers will be found in renal cancers [9, 10]. We have specifically studied MN/CA9, cadherin-6, vimentin, mucin1, pavalbumin, KIT, S100A1 in the major subtypes of renal tumors. Their expression is presented in table 1.

MN/CA9 is a promising gene marker for renal cancer. MN/CA9 is aberrantly expressed in the tumor cells but not in normal renal cells [11-13]. This antigen was first found on the cell surface of a cervical carcinoma cell line HeLa [14]. The MN/CA9 antigen was recently explored in RCC as a tumor marker. A recent immunohistochemical study characterized the antigen expression in RCC, which showed that 99% of clear cell, 65.5% granular cell, and 76.5% of mixed cell carcinomas expressed the MN/CA9 [15]. The MN/CA9 was also proved to be G250 [16].

Another potential gene marker for RCC is cadherin-6. Cadherin-6 is considered as renal tissue specific [17]. Cadherin-6 is an important gene during fetal kidney development [18, 19]. During embryogenesis, the newly formed epithelium of renal vesicle expresses E-cadherin near the ureteric bud tips and cadherin-6 more distally. In the S-shape bodies, the cadherin expression patterns reflect the developmental fate of each region. The proximal tubule progenitors express cadherin-6 while the distal tubules express E-cadherin. The role cadherin-6 and E-cadherin in kidney development has been illustrated in animal experiments. Based on these findings, the normal adult kidney possesses measurable E-cadherin and cadherin-6. Therefore, Cadherin-6 is found to be expressed in renal tissue and RCC.

DETECTION OF MN/CA9 MRNA IN RENAL CYTOPONCTION

The usefulness of fine needle aspiration (FNA) biopsy for renal tumors has been reviewed [20]. The major problems for a renal FNA biopsy are the lack of sensitivity and the difficulty in interpreting the cytological appearance. These problems often lead a cytological diagnosis to a false negative or uncertainty reported as suspect, which hinders the clinical decision. In the past few decades, no improvements have been made in differential diagnosis of renal tumors by the FNA biopsy, especially in imaging indeterminate renal lesions, which suggests that a more sensitive and objective approach should be explored.

We used the RT-PCR technique to detect the MN/CA9 gene in renal FNA biopsy [21]. A total of 35 patients with an imaging-indeterminate solid renal mass entered into this study. The molecular protocol consisted of a rapid column extraction of RNA and one-step RT-PCR for the detection of MN/CA9 gene expression. The preoperative molecular diagnosis was compared with postoperative pathology. There were 28 RCCs (19 clear cell carcinomas, 7 papillary carcinomas and 2 chromophobe carcinomas) and 7 benign tumors proved by postoperative pathology. The overall sensitivity and specificity for MN/CA9 were respectively 68% and 100%. MN/CA9 was positive in 16/19 (84%) FNA biopsies of clear cell RCCs. No false positive appeared for MN/CA9 gene expression. Moreover, MN/CA9 gene expression was positive in 8/13 (62%) of false negative or suspect cytology.

The overall sensitivity of MN/CA9 gene expression in this study was about 68%. However, 9 negative biopsies for MN/CA9 were found: 3 were clear cell carcinomas, 4 were papillary carcinomas and 2 were chromophobe carcinomas. Postoperative assays of MN/CA9 showed that the 4 papillary and 2 chromophobe carcinomas were indeed MN/CA9 negative. Among the three clear cell RCCs, two were postoperatively MN/CA9 positive and another was postoperatively negative for MN/CA9. Since our technique was highly sensitive and reproducible, we attributed the two false negatives of clear cell RCC to a possible FNA biopsy failure. No false positive was found for MN/CA9 gene expression, which supports the high specificity of the molecular assay. In our experience, the positive expression of MN/CA9 strongly suggested the malignant tumors.

This pilot study demonstrates not only that the detection of the MN/CA9 gene is feasible but also that it increases the sensitivity of renal FNA biopsy. The use of gene expression in biopsy may provide a new way to determine the malignancy.

DETECTION OF CADHERIN-6 AND MN/CA9 MRNA IN BLOOD OF RENAL TUMOR PATIENTS

The ideal diagnosis of a cancer would be a sensitive and specific blood test. During the past a few years, detection of circulating tumor cells among various solid tumors has become a focus of study. This is due to two major advances in cancer research: one is the advent of using molecular methods and the other is advances on the molecular markers available for tumor cells. With effective radical surgery of cancer, the outcome of patients is more and more linked with early tumor cell dissemination. At present, the decision as to whether adjuvant therapy should be applied to prevent metastatic relapse is based on the basis of traditional staging system (TNM classification) instead of on the individual patient's risk to develop recurrence. It is argued that TNM system could result in failure of the primary treatment of an individual patient because residual micrometastasis tumor cells may increase the individual risk to develop tumor recurrence. Therefore, the detection of micrometastatic cancer cells may improve the staging system.

We designed and developed the standard and nested RT-PCR by using MN/CA9 and cadherin-6 [22]. By the standard RT-PCR, we can easily detect one tumor cell/ 5 x 10[5] PBMNCs. By using cadherin-6, we detected the circulating renal cancer cells. We could find that 2/17 (11.7%) normal blood samples were positive. For RCC patients, cadherin-6 mRNA could be found positive in 16/31 (51.6%) blood samples. 6 (75%) were cadherin-6 positive among 8 blood samples from metastasized patients (M+ or N+). This study demonstrates that cadherin-6 is a feasible marker to detect the renal cancer cells in blood. We also confirmed that the circulating renal cancer cells could be detected by MN/CA9. Our preliminary impression was that cadherin-6 is easier to find positive than MN/CA9. In fact, we should use a nested RT-PCR of MN/CA9 to enhance the sensitivity and then we can find a clear signal (figure 2). We think that the selection of mRNA markers can affect the detection sensitivity.

Two groups have published some preliminary results on detection of circulating renal cancer cells by using MN/CA9 [23, 24].

Detection of S100A1 Gene in Renal Tumor Samples for the Subtype Differentiation

The differentiation between the granular (eosinophilic) variant of chromophobe RCC and oncocytoma may be difficult because the morphological features overlap [25 – 30]. Typically, renal oncocytoma is composed exclusively of oncocytes. The oncocytes are the large eosinophilic granular cells with uniform and round nucli and inconspicuous or absent nucleoli, arranged in distinct nest separated by fibrovascular septa. The chromophobe RCC may also have the eosinophilic cells, which cause diagnostic problem with the oncocytoma.

S100A1 gene belongs to at least 10 S100 gene family on human chromosome 1q21. S100 proteins were reported to be involved in different biological activities such as transduction of intracellular calcium signaling and cytoskeleton-mediated interactions as well as cell cycle progression and cell differentiation. S100A1 is preferentially studied in cardiac tissues [31].

We have tested the S100A1 gene expression as a marker to differentiate the chrompophobe RCC from renal oncocytoma [32, 33]. We have studied 39 tumor samples (9 clear cell RCCs, 6 papillary RCCs, 9 chromophobe RCCs and 15 oncocytomas). The gene expressions of S100A1 was analyzed by RT-PCR. In total, 14/15 (93%) of oncocytomas expressed the S100A1 gene. On the contrary, all 9 chromophobe RCCs presented no S100A1 gene and considered as negative. There was a significant difference in the positive percentages of S100A1 gene expression between these two subtypes ($p < 0.01$, the Fisher exact test). The protein expression of S100A1 was evaluated by immunohistochemistry. 9 oncocytomas showed a strong staining of S100A1. 4 oncocytomas were scored as moderate staining and one as weak staining. The S100A1 staining was absent in all 9 chromophobe RCCs. S100A1 immunohistochemical staining was also observed in 6/9 clear cell and 4/6 papillary carcinomas. The results of S100A1 gene expression corresponded well with the results of immunohistochemical staining.

Since there is a significant difference in S100A1 gene expression between the chromophobe RCC and the oncocytoma, S100A1 as a gene marker should be used to distinguish these two subtypes of renal tumors. Very recently, Lin et al reported S100 in renal tumors [34]. They used a polyclonal antibody against S100. They found that 1 of 16 chromophobe RCCs expressed S100 while 13 (87%) of 15 oncocytomas expressed S100. The utilization of S100A1 as a marker to the differentiation may be more specific than that of S100.

New techniques such as flow cytometry, fluorescence in situ hybridization or comparative genomic hybridization have been recently studied to differentiate the subtypes of renal tumors by detecting chromosome aberrations [35, 36, 37], but these techniques are difficult to perform. The technique of RT-PCR for gene expression has been well developed for the utilization in surgical pathology of solid tumors [38]. Therefore, the RT-PCR can be used to provide a confirmatory result. Our data indicate that S100A1 may be used as a potential marker for the differential diagnosis between the chromophobe RCC and oncocytoma.

Conclusion

Some mRNA or gene markers are available for detection of renal cancer cells. The detection of mRNA expression can be used as an adjuvant method for the diagnosis and the subtype differentiation of renal tumors.

Acknowledgement

We thank la ligue nationale contre le cancer du comité de la Loire, French Ministry of Health (PHRC) and INCA for providing the financial support. We also thank Dr Oosterwijk E for providing us with renal carcinoma cell lines.

References

[1] Li, G.; Cuilleron, M.; Gentil-Perret, A.; Tostain, J. Characteristics of image-detected solid renal masses : implication for optimal treatment.. *Int. J. Urol.* 2004, 11, 63-67.

[2] Lane, B.R.; Kattan, M.W. Predicting outcomes in renal cell carcinoma. *Curr. Opin. Urol.* 2005, 15, 289-297.

[3] Li, G.; Cuilleron, M.; Cottier, M.; Gentil-Perret, A.; Passebosc-Faure, K.; Lambert, C.; Genin, C.; Tostain, J. Rapid and sensitive detection of messenger RNA expression for molecular differential diagnosis of renal cell carcinoma. *Clin. Cancer Res.* 2003, 9, 6441-6446.

[4] Boylan, S.; Honda, S.; Hjelmeland, L.M.; Handa, J. T. An optimized protocol for first strand cDNA synthesis from laser capture microdissected tissues. *Lab. Invest.* 2001, 8, 1167-1169.

[5] Luzzi, V.; Holtschlag, V.; Watson, M.A. Expression profiling of ductal carcinoma in situ by laser capture microdissection and high density oligonucleotide arrays. *Am. J. Pathol.* 2001, 158, 2005-2010.

[6] Zambrano, N.R.; Lubensky, I.A.; Merino, M.J.; Linehan, W.M.; Walther, M.M. Histopathology and molecular genetics of renal tumors: toward unification of a classification system. *J. Urol.* 1999, 162: 1246-1258.

[7] Cheville, J.C.; Lohse, C.M.; Zincke, H.; Weaver, A.L.; Blute, M.L. Comparisons of outcome and prognostic features among histologic subtypes of renal cell carcinoma. *Am. J. Surg. Pathol.* 2003, 27, 612-624.

[8] Li, G.; Passebosc-Faure, K.; Lambert, C.; Gentil-Perret, A.; Blanc, F.; Oosterwijk, E.; Genin, C.; Tostain, J. Flow cytometric analysis of antigen expression in malignant and normal renal cells. *Anticancer Res.* 2000, 20, 2773-2778.

[9] Higgins, J.P.T. ; Shinghal, R. ; Gill, H. ; Reese, J.H. ; Terris, M. ; Cohen, R.J. Fero, M. ; Pollack, J.R. ; Van De Rijn, M. ; Brooks, J.D. Gene expression paterns in renal cell carcinoma assessed by complementary DNA microarray. *Am. J. Pathol.* 2003, 162, 925-932.

[10] Moch, H. ; Schrami. ; Bubendorf, L. ; Mirlacher, M. : Kononen, J. ; Gasser, T. ; Mihatsch, M.J. ; Kallioniemi, O.P. ; Guido, S. High throughout tissue microarray anlysis to evaluate genes uncovered by cDNA microarray screening in renal cell carcinoma. *Am. J. Pathol.* 1999, 154, 981-986.

[11] McKiernan, J.M. ; Buttyan, R. ; Bander, N.H. ; Stifelman, M.D. ; Katz, A.E. ; Chen, M. ; Olsson, C.A ; Sawczuk, I.S. Expression of the tumor associated gene MN: a potential for human renal cell carcinoma. *Cancer Res.* 1997, 57, 2362-2365.

[12] Li, G. ; Passebosc-Faure, K. ; Lambert, C. ; Gentil-Perret, A. ; Blanc, F. ; Oosterwijk, E. ; Genin , C. ; Tostain, J. The expression of G250/MN/CA9 antigen by flow cytometry : its possible implication for detection of micrometastatic renal cancer cells. *Clin. Cancer Res.* 2001, 7, 89-92.

[13] Liao, S.Y. ; Aurelio, O.N. ; Jan, K. ; Zavada. J, ; Stanbridge, E.J. Identification of the MN/CA9 protein as a reliable diagnostic biomarker of clear cell carcinoma of the kidney. *Cancer Res.* 1997, 57, 2827-2833.

[14] Pastorek, J. ; Pastorekkova, S. ; Callbaut, I. ; Mournon, J. ; Mo,.P. ; Zelnik, V. ; Opavsky, R. ; Zativicova, M. ; Liao, S. ; Portell, D. ; Stanbridge, E.J. ; Zavada, J. ; Burny, A. ; Kettman, R. Cloning and characterization of MN, a human tumor-associated protein with a domain homologous to carbonic anhydrase and putative helix-loop-helixDNA binding segment. *Oncogene.* 1994, 9, 2877-2887.

[15] Uemura, H. ; Nakagawa, Y. ; Yoshida, K. ; Saga, S. ; Yoshikawa, K. ; Hirao, Y. ; Oosterwijk, E. MN/CA IX/ G250 as a potential target for immunotherapy of renal cell carcinomas. *Brit. J. Cancer.* 1999, 81, 741-746.

[16] Grabmaier, K. ; Vissers, J. L. ; DeWeijert, M. C. ; Oosterwijk-Wakka, J. C. ; Van Bokhoven, A. ; Brakenhoff, R. H. ; Noessner, E. ; Mulders, P.A. ; Merkx, G. ; Figdor, C.G. ; Adema, G.J. ; Oosterwijk, E. Molecular cloning and immunogenicity of renal cell carcinoma-associated antigen G250. *Int. J. Cancer.* 2000, 85, 865-870.

[17] Paul, R. ; Ewing, C.M. ; Robinson, J.C. ; Marshall, F.F. ; Johnson, K.R. ; Wheelock, M.J. ; Issacs, W.B. Cadherin-6, a cell adhesion molecule specifically expressed in the proximal renal tubule and renal cell carcinoma. *Cancer Res.* 1997, 57, 2741-2748.

[18] Cho, E. A. ; Patterson, L. T. ; Brookhiser, W. T. ; Mah, S. ; Kintner, C. ; Dressler, G. R.:Differential expression and function of cadherin-6 during renal epithelium development. *Development.* 1998, 125, 803-812.

[19] Shimazui, T. ; Oosterwijk, E. ; Akaza, H. ; Bringuier, P. ; Ruijter, E. ; van Berkel, H. ; Wakka Oosterwijk, J. ; van Bokhoven, A. ; Debruyne, F. M. J. ; Schalen, J. A. Expression of cadherin-6 as a novel diagnostic tool to predict prognosis of patients with E-cadherin absent renal cell carcinoma. *Clin. Cancer Res.* 1998, 4, 2419-2424.

[20] Renshaw, A. ; Granter, S.R. ; Cibas, E.S. Fine-needle aspiration of the adult kidney. *Cancer.* 1997, 81, 71-88.

[21] Li, G. ; Cuilleron, M. ; Cottier, M. ; Gentil-Perret, A. ; Lambert, C. ; Genin, C. ; Tostain, J. The use of MN/CA9 gene expression in identifying malignant solid renal tumors. *Eur. Urol.* 2006, 49, 401-405.

[22] Li, G. ; Passebosc-Faure, K. ; Gentil-Perret, A. ; Lambert, C. ; Genin, C. ; Tostain, J. Cadherin-6 gene expression in conventional renal cell carcinoma: a useful marker to detect circulating tumor cells. *Anticancer Res.* 2005, 25, 377-382.

[23] McKiernan, JM. ; Buttyan, R. ; de le Taille, A. ; Stifeman, M.D. ; Emanuel, E.R.. ; Bagiella, E . ; Rubin, A.; Katz, A.. ; Olsson, C.A. ; Sawczuk, I.S. The detection of renal carcinoma cells in the peripheral blood with an enchanced reverse transcriptase polymerase chain reaction assay for MN/CA9. *Cancer.* 1999, 86, 492-497.

[24] de la Taille, A. ; Katz, A. ; Cao, Y. ; McKiernan, J. ; Buttyan, R. ; Burchardt, M. ; Burchardt, T0. ; Hayek, O. ; Olsson, C.A. ; Chopin, D.K. ; Sawczuk, I.S. Blood-based RT-PCR assays of MN/CA9 or PSMA: clinical application in renal cancer patients. *Urology.* 2000, 56, 393-398.

[25] Mete, O.; Kilicaslan, I.; Gulluoglu, M.G.; Uysal, V. Can renal oncocytoma be differentiated from its renal mimics? The utility of antimitochondrial, caveolin 1, CD63 and cytokeratin 14 antibodies in the differential diagnosis. *Vichows Arch.* 2005, 447, 938-946.

[26] Skinnider, B.F.; Folpe, A.L.; Hennigar, R.A. *et al.* Distribution of cytokeratins and vimentin in adult renal neoplasms and normal renal tissue: potential utility of a cytokeratin antibody panel in the differential diagnosis of renal tumors. *Am. J. Surg. Pathol.* 2005, 29; 747-754.

[27] Mazal, P.R.; Exner, M.; Haitel, A.; Krieger, S.; Thomson, R.B.; Aronson, P.S.; Susani, M. Expression of kidney-specific cadherin distinguishes chromophobe renal cell carcinoma from renal oncocytoma. *Hum. Pathol.* 2005, 36, 22-28.

[28] Shen, S.S.; Krishna, B.; Chirala, R.; Amato, R.J.; Troung, L.D. Kidney-specific cadherin, a specific marker for the distal portion of the nephron and related renal neoplasms. *Mod. pathol.* 2005, 18, 933-940.

[29] Brunelli, M.; Eble, J.N.; Zhang, S.; Martignoni, G,; Delahunt, B.; Cheng, L. Eosinophilic and classic chromophobe renal cell carcinoma have similar frequent losses of multiple chromosomes from among chromosomes 1, 2, 6, 10, and 17, and this pattern of genetic abnormality is not present in renal oncocytoma. *Mod. Pathol.* 2005, 18, 161-169.

[30] Patton, K.T.; Tretiakova, M.S.; Yao, .JL.; et al. Expression of RON proto-oncogene in renal oncocytoma and chromophobe renal cell carcinoma. *Am. J. Surg. Pathol.* 2004, 28, 1045-1050.

[31] Most, P.; Bernotat, J.; Ehlermann, P.;e et al. S100A1: a regulator of myocardial contractility. *PANS* 2001, 98, 13889-13894.

[32] Li, G. ; Gentil-Perret, A. ; Lambert, C. ; Genin, C. ; Tostain, J. S100A1 and KIT gene expressions in common subtypes of renal tumors. *Eur. J. Sur. Oncol.* 2005, 31 , 299-303.

[33] Li, G. ; Barthelemy, A. ; Feng, G. ; Gentil-Perret, A. ; Peoc'h, M. ; Genin, C. ; Tostain, J. S100A1 : *a powerful marker to differentiate chromophobe renal cell carcinoma from renal oncocytoma.* Histopathology in press.

[34] Lin, F.; Yang, W.; Betten, M.; Teh, B.T.; Yang, X.J. The french kidney cancer study group. Expression of S100 protein in renal cell neoplasms. *Hum. Pathol.* 2006, 37, 462-470.

[35] Li, G.; Cottier, M.; Sabido, O.; et al. Different DNA ploidy patterns for the differentiation of common subtypes of renal tumors. *Cell Oncol.* 2005, 27, 51-56.

[36] Heinze, F.; Kovacs, G. Identifying BAC clones for diagnosis of conventional renal cell carcinoma by FISH. *Histopathology.* 2002, 4, 308-312.

[37] Wilhelm, M.; Veltman, J.A.; Olshen AB.; et al. Array-based comparative genomic hybridization for the differential diagnosis of renal cell carcinoma. *Cancer Res.* 2002, 62, 957-960.

[38] Raj, G. V. ; Moreno, J. G. ; Gomella, L. G. Utilization of polymerase chain reaction technology in the detection of solid tumors. *Cancer.* 1998, 82, 1419-1442.

In: Messenger RNA Research Perspectives
Editor: Takeo Takeyama, pp. 125 -144
ISBN: 978-1-60021-814-9

Chapter V

Canonical and Non Canonical Poly(A) Polymerases: From Human to the Protozoan Parasite *Entamoeba histolytica*

***César López-Camarillo*[*,1], *Laurence A. Marchat*[2], *Esther Orozco*[3], *Elisa Azuara Liceaga*[1], and *Jessica García-Vivas*[2]**

[1] Posgrado en Ciencias Genómicas, Universidad Autónoma de la Ciudad de México, Av. San Lorenzo # 290. Col del Valle, CP 03100 México D.F., México.

[2] Programa Institucional de Biomedicina Molecular, Escuela Nacional de Medicina y Homeopatía-IPN, Guillermo Massieu Helguera #239, Fracc. La Escalera Ticomán, CP 07320 México D.F., México.

[3] Departamento de Patología Experimental, CINVESTAV-IPN, Av. IPN # 2508, Col. San Pedro Zacatenco, CP 07360 México D.F., México.

Abstract

The pre-mRNA polyadenylation in the nucleus is a critical event in eukaryotic gene expression since it is involved in messengers transport, stability and translation regulation. The enzyme responsible for the poly(A) tail addition is the canonical poly(A) polymerase that belongs to the DNA polymerases β-like nucleotidyltransferases superfamily. Recently, a distinct polyadenylation reaction has been described for certain transcripts involved in growth, cell cycle, development, genome maintenance, synaptic plasticity and long-term memory storage. This reaction is catalyzed by a different family of poly(A) polymerases identified as TRF4, GLD-2 and Cid proteins in *Homo sapiens, Xenopus laevis, Caenorhabditis elegans, Saccharomyces cerevisiae* and *Schizosaccharomyces pombe*.

In *Entamoeba histolytica,* the protozoan parasite responsible for human amoebiasis, we have previously reported the identification of the pre-mRNA 3´end cleavage and polyadenylation machinery. We have characterized the poly(A) polymerase EhPAP,

[*] Corresponding author: Dr. César López-Camarillo, Email: genomicas@yahoo.com.mx, Tel: (+55) 5488 6661 ext 5352.

which shares high sequence homology with canonical nuclear PAPs from other species. Here, we showed that recombinant EhPAP was able to bind RNA, even when it does not have a conserved RNA binding domain. Next, we extended our studies on mRNA polyadenylation and described a novel family of PAPs in this pathogen. By genomic survey and *in silico* analysis we identified six TRF4/GLD-2/Cid orthologous proteins in *E. histolytica.* Predicted proteins sequences are highly divergent from classical eukaryotic PAPs, but they possess the hallmarks of the DNA polymerases β-like nucleotidyltransferases superfamily, including the amino acids involved in nucleotide interaction and catalysis. *E. histolytica* non canonical PAPs have no apparent RNA recognition motif, which is critical for RNA binding. By RT-PCR assays we showed that genes are expressed in *in vitro* cultured trophozoites but with differential expression levels. Finally, phylogenetic and PSI-BLAST analysis showed that the PAP central domain of conventional and non conventional PAPs appears to have emerged from a bacterial ancestor known as the minimal nucleotidyltransferase (MNT) domain. The striking diversity of poly(A)-adding enzymes in *E. histolytica* suggests a greater degree of control over poly(A) tail length and mRNA stability than it has already been thought in this protozoan parasite. The further functional characterization of *E. histolytica* TRF4/GLD-2/Cid orthologous proteins would provide new insights into the biological roles of polyadenylation and mRNA turnover in this pathogen.

1. Introduction

Eukaryotic gene expression involves several events from the transcription of a gene by RNA Polymerase II (RNA Pol II) in the nucleus to protein synthesis in the cytoplasm. Messenger RNA precursors (pre-mRNAs) are modified through a serie of nuclear processing reactions, which include the addition of a 7-methylguanosine cap at the 5′ end (capping), introns removal (splicing) and cleavage/polyadenylation of 3′ untranslated region (3′ UTR). Then, mature RNA molecules can be exported to the cytoplasm through the nuclear pore complex to be translated. Proteins synthesis depends on mRNA turnover and translational efficiency, which can be modulated through poly(A) tail removal or addition. In particular, mRNA polyadenylation in the cytoplasm provides cells with an additional mechanism of gene expression regulation that is important for transcripts participating in early development, cell growth and synaptic plasticity. Unconventional RNA polyadenylation is also involved in the nuclear RNA surveillance pathway. Thus, RNA polyadenylation is considered as a global regulatory event in eukaryotic gene expression. Here we present a general description of the mechanisms and proteins involved in RNA polyadenylation in eukaryotic cells. We have also included our own results about the poly(A) polymerases (PAP) family in the protozoan parasite Entamoeba histolytica. We identified a single canonical EhPAP and six non canonical PAPs, which all possess the hallmarks of the DNA polymerases β-like nucleotidyltransferases superfamily. The diversity of PAPs in E. histolytica indicates the relevance of poly(A) tail length and RNA stability for gene expression regulation in this pathogen.

2. Nuclear Pre-mRNA Polyadenylation in Human and Yeast

Numerous studies have shown that the poly(A) tail at mRNA 3′ end is a central element for gene expression regulation. First, 3′ end cleavage and polyadenylation promotes transcription termination [1] and nuclear export of transcripts [2]. The poly(A) tail also contributes to translation efficiency through interaction with the Poly(A) Binding Protein (PAB) bound to the poly(A) tract, the translation initiation factor elF-4G and several mRNA cap-binding proteins [3; 4]. Moreover, the poly(A) tail is involved in mRNA stability, since its elimination by poly(A) ribonucleases is the first step in the main mRNA degradation pathways [5].

Transcripts polyadenylation in the nucleus is a multistep reaction that involves pre-mRNA 3′ end cleavage followed by adenosine residues addition. In human, the Cleavage and Polyadenylation Specificity Factor (CPSF, four subunits) recognizes the consensus polyadenylation signal (AAUAAA) [6]. It interacts with FIP1 and the Poly(A) Polymerase (PAP) to tether it to RNA [7]. The trimeric Cleavage Stimulating Factor (CstF) binds to the U-rich downstream element and the CPSF major subunit, enhancing its RNA binding capacity. Then, the Cleavage Factors (CFIm and CFIIm) interact with the polyadenylation site to perform RNA cleavage. Finally, PAP catalyzes the poly(A) tail synthesis in a processive reaction that involves CPSF, FIP1, CFIm and PABII [8]. The mRNA 3′ end processing machinery includes additional factors, such as symplekin, Rrbp6, PNAS-120 and PC4. Homologues of cleavage/polyadenylation factors have been identified in Saccharomyces cerevisiae and other species, suggesting that mRNA 3′ end processing is conserved through evolutionary scale [9].

The 3′ end formation is not an independent mRNA processing event, but it shares close relationships with other pre-mRNA modifying reactions. The PAP C-terminal domain interacts with the U2AF65 factor, establishing a functional link between polyadenylation and pre-mRNA splicing. The Cap Binding Complex (CBC) interacts with CPSF, stabilizing the polyadenylation complex. Similarly, CBC enhanced the interaction between U1 snRNP and the 5′ end of the first intron [9]. Pre-mRNA processing reactions are also tightly linked with transcription. In particular, PC4, CPSF and CstF interact with RNA Pol II C-terminal domain and TATA-Box Binding Protein (TBP) to promote transcription initiation. These interactions are maintained during RNA elongation, and transcription termination depends on the binding of CPSF and CstF to poly(A) signals at the pre-mRNA 3’ UTR [9]. Interestingly, pre-mRNA 3′ end formation is also linked to cell cycle progression. In Mammals, fluctuations in polyadenylated mRNAs synthesis and transcripts utilization through cell cycle are consistent with variations in the amount of CstF and CF complexes, with an increase in S phase and a general repression of gene expression in mitosis. In addition, PAP is inactivated in mitosis through hyperphosphorylation by the $p34^{cdc2}$-cyclin B complex, resulting in a reduction of polyadenylated transcripts and proteins synthesis [10; 11].

The poly(A) tail formation in the nucleus is catalyzed by the canonical nuclear PAP that belongs to the DNA polymerase β-like nucleotidyltransferase superfamily. PAPs are highly conserved through evolutionary scale; they present a tripartite domain organization that includes the catalytic domain with three invariant aspartate residues involved in nucleotide transfer and a F/YGS motif responsible for ATP binding. At the C-terminal end, they have a low conserved RNA-Binding Domain (RBD) [12].

3. Non Canonical Poly(A) Polymerases in Human, Frog and Yeast

3.1. Molecular Organization and Classification

A distinct family of PAPs, knowns as TRF4/GLD-2/Cid proteins, was recently characterized in eukaryotic cells. Although they do not share high sequence homology with canonical PAPs, these unconventional PAPs have the conserved central domain with the specific aminoacids for ATP binding and catalytic activity. However, they lack the C-terminal RBD and are therefore inactive on their own. It has been demonstrated that Caenorhabditis elegans GLD-2 interacts with GLD-3 [13] (Airlp/Air2p in yeast [14]), a developmental regulator of the bicaudal-C family of RNA binding proteins [15] that provides the RBD and reconstitutes the tripartite domain organization of canonical PAPs to restore catalytic activity.

Experimental data and sequences homology from BLAST analysis allowed the separation of two groups of non canonical PAPs: i) GLD-2 homologues include GLD-2 from Homo sapiens (also called PAPD4) [14], Xenopus laevis [16; 17] and C. elegans [13], as well as Cid1, Cid11, Cid13 and Cid16 from S. pombe [18]; whereas ii) TRF4 homologues include H. sapiens TRF4-1 (POLS) and TRF4-2 (PAPD5) ([19], Schizosaccharomyces pombe Cid12 and Cid14 [18], as well as S. cerevisiae TRF4 (PAP2 or POLσ) and TRF5 [19, 20]. Whereas most GLD-2 homologues have been found in the cytoplasm, TRF4 and TRF5 proteins are nuclear enzymes [21], indicating that non classical PAPs may be involved in distinct biological processes.

3.2. Molecular Events of the Cytoplasmic Polyadenylation

The cytoplasmic polyadenylation reaction has been mainly studied in Xenopus, Mus musculus and Drosophila development [22]. It involves the participation of the polyadenylation signal (AAUAAA), which is also essential for nuclear polyadenylation, and a U-rich sequence called the Cytoplasmic Polyadenylation Element (CPE). The trans-acting factors are: i) the Cytoplasmic Polyadenylation Element Binding protein (CPEB), a zinc-finger protein that contains an RNA Recognition Motif (RRM); ii) a cytoplasmic form of CPSF and iii) a cytoplasmic non conventional PAP. In Xenopus resting oocytes, CPEB is bound to CPE. When oocytes enter in meiosis, progesterone stimulation leads to a transient decrease in cAMP and the activation of Eg2, a member of the aurora family of protein kinases. Eg2 phosphorylates CPEB, enhancing its affinity to CPE and CPSF-160 subunit. This interaction stabilizes CPSF bound to the polyadenylation signal, inducing the recruitment of GLD-2 to initiate the poly(A) tail elongation [23]. Recently, the CstF-77 subunit was identified in punctuate cytoplasmic foci in association with elF4E, CPSF-100 subunit, CPBE and GLD-2, suggesting that it is also involved in cytoplasmic polyadenylation [24].

Although they share some characteristics, nuclear and cytoplasmic polyadenylation reactions are readily distinct: i) the cytoplasmic CPSF lacks the 70 kDa subunit [25; 26]; ii) the cytoplasmic reaction is not associated with any cleavage reaction; iii) it is restricted to a few species of transcripts containing the CPE; and finally, iv) the cytoplasmic PAP is activated in mitosis, whereas the nuclear PAP is inhibited in this phase [10].

3.3. Biological Significance of Non Canonical Polyadenylation

GLD-2 proteins were initially associated with translational activation of certain mRNAs in cells with specific requirements, such as germ cells. Xenopus prophase-I arrested oocytes contain a small subset of mRNA that is partially deadenylated and stored in an inactive form in the cytoplasm [27]. When meiosis process is stimulated, cytoplasmic readenylation by GLD-2 promotes translational activation of dormant mRNAs that are required for cell development [28].

Recently, cytoplasmic poly(A) polymerases have also been related to brain functions. In neuronal cells, stimulation of synapses activates cytoplasmic polyadenylation to synthesize proteins that are important for long-term potentiation and memory storage [29; 30]. The abundance of GLD-2 mRNA varies in the different parts of the brain, suggesting the existence of a regional specificity of cytoplasmic polyadenylation [17].

In unicellular eucakyotic cells, cytoplasmic polyadenylation is important for DNA replication and genome maintenance through cell cycle. In S. pombe, Cid13 specifically polyadenylates the transcript of Succ22, an enzyme involved in dNTPs synthesis. This reaction bypasses the classical steps of transcription and translation, insuring rapid and optimum dNTPs availability for cells growing in unfavorable conditions [31]. Cid1 acts throughout the checkpoint-signaling Chk1 to inhibit unscheduled mitosis when DNA polymerase δ or ε are inactivated and there is no elongation of RNA primers during replication [32].

S. cerevisiae TRF-4 and TRF-5 were first described as nuclear DNA polymerases required for sister chromatides cohesion [33; 34]. Then it has been shown that TRF-4 is a nuclear poly(A) polymerase [35] that plays a role in 3′ end processing of noncoding RNA precursors, such as pre-snRNA [36] and pre-rRNA [37], demonstrating that polyadenylation is not restricted to mRNAs. TRF4/TRF5 form a multiprotein complex with Airp1/Airp2 and the RNA helicase Mtr4p to polyadenylate aberrant rRNA and tRNA and mediate their subsequent degradation by the nuclear exosome [38-40].

Unconventional RNA polyadenylation has been studied in a very limited number of species. All the available data strongly indicate that this event is an essential control point for gene expression regulation. However, there is no information about this process in the protozoan organisms that cause the main parasitic diseases, including Entamoeba histolytica.

4. THE ENTAMOEBA HISTOLYTICA POLY(A) POLYMERASES FAMILY

Entamoeba histolytica is the early branching protozoan responsible for human amoebiasis that is characterized by dysentery and liver abscess. This disease is related to bad sanitary conditions in many tropical developing countries. It affects about 500 millions people worldwide causing substantial morbidity and mortality that contribute to the low economic development of these regions [41]. Despite its medical importance, little is known about the control of gene expression in this parasite. To date, most studies on drug resistance, pathogenesis or virulence genes have focused on transcriptional level and several transcription factors have been cloned and characterized [42-44]. However, the posttranscriptional regulation mechanisms including capping, splicing and polyadenylation reactions, remain almost unknown. We have recently reported that changes in poly(A) tail length regulate half-life and stability of EhPgp5 mRNA, demonstrating the relevance of posttranscriptional regulation for the multidrug resistance (MDR) phenotype of E. histolytica [45]. The recent publication of the whole genome sequence of E. histolytica [46] and the availability of this large raw data set should rapidly allow great progress in the identification and characterization of genes, proteins and biochemical pathways that are important for parasite survival, as well as the dilucidation of molecular machineries that regulate gene expression. Taking advantage of the E. histolytica genome sequences available in TIGR (http://www.tigr.org/tdb/e2k1/eha1/), Sanger (http://www.sanger.ac.uk/Projects/E_histolytica/) and Pathema (http://pathema.tigr.org/tigr-scripts/Entamoeba/PathemaHomePage.cgi) databases, we used mammalian and yeast proteins sequences to search for genes coding for proteins that could be involved in poly(A) tail formation in this parasite [47]. We identified the four subunits of EhCPSF (160, 100, 73 and 30 kDa) as well as EhFIP1 and EhPAP, the three subunits of EhCstF (77, 64 and 50 kDa), the two subunits of EhCFIIm (EhClP1 and EhPCF11) but only the 25 kDa subunit of the EhCFIm. We also identified additional factors, such as EhMpe1, EhSsu72 and EhPC4/Sub1. In addition, multiple alignment of genomic and cDNA 3′ ends made possible the detection of putative poly(A) signals that may be bound by these trans-acting factors: the polyadenylation signal UA(A/U)UU previously reported by Bruchhaus et al. [48] and an U-rich motif downstream the poly(A) site. Our results suggested that E. histolytica has an evolutionary conserved pre-mRNA 3′ end cleavage and polyadenylation machinery. Based on the knowledge of pre-mRNA 3' end processing in human and yeast, and considering that this process is conserved through the evolutive scale, we proposed a hypothetical working model for pre-mRNA cleavage and polyadenylation reaction in E. histolytica. EhCPSF complex could bind to the polyadenylation signal UA(A/U)UU; EhCstF complex could recognize the U-rich motif and recruit EhCFIm-25, EhClp1 and EhPcf11 proteins to perform RNA cleavage at the poly(A) site. Finally, EhPAP could synthesize the poly(A) tail, in a reaction stimulated by the interaction with other processing factors. Additional interactions with EhMpe1, EhPC4/Sub1, EhSsu72, and EhPfs2 factors, could also contribute to the efficient cleavage and polyadenylation reaction of pre-mRNA 3′ end in this parasite [47].

4.1. The Canonical Poly(A) Polymerase EhPAP of *E. histolytica*

In order to characterize mRNA 3′ end formation in E. histolytica, we first focused on the enzyme responsible for the poly(A) tract elongation in the nucleus. We recently identified a single gene that codes for a putative canonical poly(A) polymerase (EhPAP) of 522 amino acids (aa) that presents high identity (25-32%) and homology (43-52%) with yeast, protozoa, nematodes, plants and mammals PAPs through BLASTP analysis [49]. Bioinformatics studies showed that EhPAP presents most characteristics of the canonical PAPs. It has the highly conserved N-terminus that contains the central catalytic domain (36 to 131 aa) with the three invariant Aspartate residues (D_{76}, D_{78} and D_{130}) involved in nucleotide transfer and the F/YGS motif (63 to 65 aa) responsible for ATP binding in homologous enzymes. At the less conserved C-terminus, we identified a RBD region (322 to 444 aa) that is divergent from the one described in homologous proteins, as well as several phosphorylation sites, which could participate in the poly(A) polymerase activity regulation. EhPAP does not have a typical nuclear localization signal, suggesting that E. histolytica could use other sequence signals for the translocation of this protein to the nucleus. All these signature sequences showed that EhPAP is a member of the polymerases β-like superfamily of nucleotidyltransferases [49].

The EhPap gene was amplified by PCR using specific oligonucleotides and genomic DNA from clone A trophozoites (strain HM1: IMSS) and cloned into the pRSET-A expression vector. The recombinant rEhPAP protein expressed in IPTG induced Escherichia coli bacteria were used to generate polyclonal antibodies in mouse. In Western blot assays, anti-EhPAP antibodies recognized a 60 kDa protein in trophozoites nuclear extracts, as well as an additional polypeptide of 63 kDa, which could result from phosphorylation at the putative sites previously identified by computational analysis. The detection of EhPAP in the nucleus is in good agreement with the biological function of a factor involved in pre-mRNA poly(A) tail formation. The same bands were also detected in cytoplasmic extracts where proteins are synthesized [49].

We next performed inmunofluorescence and microscopy confocal assays to study the in situ localization of EhPAP in trophozoites of clone A. Cells were grown on sterile coverlips, fixed (4% paraformaldehyde), permeabilized (0.5% Triton-PBS) and blocked with 1% (w/v) BSA. Then, they were incubated with polyclonal antibodies against EhPAP (1:5000) and a goat anti-mouse antibody labeled with fluorescein isothiocynate (FITC) (1:2000). As negative control, we omitted the first antibody. Trophozoites were counterstained with propidium iodide (1 μg/ml) and examined through a Leica microscope attached to a laser confocal scanning system (data not shown). Our results confirmed that EhPAP is located both in the nucleus and cytoplasm of trophozoites, in agreement with studies in other organisms [50]. It is possible that the cellular location of EhPAP varies through cell cycle progression, being translocated from the cytoplasm to the nucleus, as it has been reported for other proteins involved in mRNA metabolism [51]. Interestingly, in some trophozoites, EhPAP appears to be located in punctuate nuclear foci. A similar distribution in nuclear foci has been reported for several factors involved in mRNA splicing and cleavage/polyadenylation (including PAP), suggesting that mRNA transcription and processing are concentrated at specific sites in the nucleus [52]. However, studies are required to confirm these assumptions in E. histolytica.

We have previously shown that the poly(A) tail of the EhPgp5 mRNA was increased in multidrug-resistant clone C2 trophozoites growing in higher emetine concentrations [45]. We next demonstrated by RT-PCR assays that EhPap mRNA expression is increased in clone C2 trophozoites cultured in the presence of the drug. It is possible that the activation of EhPap transcription and the subsequent increase of EhPAP synthesis allows the formation of a longer EhPgp5 mRNA poly(A) tail, increasing EhPgp5 mRNA stability and contributing, at least in part, to the generation of MDR phenotype in drug-resistant trophozoites [49]. In addition, EhPap mRNA expression varies during cell cycle progression, being predominantly detected in G1 and S phases, but not in M and G2 phases. These results suggested that EhPAP expression and functions could be regulated through cell cycle progression. It is also possible that EhPAP plays a regulatory role in this cellular process as reported in other organisms [49].

Using the bovine PAP crystal data (PDB 1F5A) [12] and the Geno3D (www.pbil.ibcp.fr) and Rasmol (www.umass.edu/microbio/rasmol/getras.htm) programs, we obtained the predicted three-dimensional structure of EhPAP (figure 1A). Like most polymerases, EhPAP is a U-shaped molecule with a tripartite domain organization. Catalytic and RNA-binding domains form a groove in which the central domain is thought to orient the incoming ATP and RNA molecule for poly(A) tail extension. The EhPAP C-terminus is similar to the one of bovine PAP; it folds into a compact globular domain consisting of three-stranded antiparallel β-sheets flanked on one side by two helices. Intriguingly, four β-sheets have been described in the bovine PAP C-terminus [12] although amino acids sequences are different between both species. EhPAP C-terminal domain shares topological similarities with RNA binding proteins that lack the conserved RNA Recognition Motif (RRM) and bind RNA through β-sheet surfaces [12]. Since we did not detect a conserved RBD in EhPAP, it is possible that it uses the β-sheets structures at the C-terminal domain to bind RNA.

To determine whether the absence of a conserved RBD affect EhPAP RNA binding capacity, we performed RNA-binding assays using the purified recombinant rEhPAP protein and the 3′UTR of the EhPgp5 gene ($PSIII^{156}$ fragment) as a [α-^{32}P] UTP-labeled RNA probe [45]. RNA-protein complexes were resolved at 130 V for 4 h on pre-electrophoresed 6% nondenaturing PAGE. Gels were vacuum-dried, and RNA-protein interactions were detected by scanning in a PhosphorImager Apparatus. rEhPAP was able to form four complexes with RNA, suggesting that the lack of a conserved RBD does not alter its binding activity. Interestingly, the number of complexes correlates with the number of putative polyadenylation/cleavage sites identified in this riboprobe, although no specific binding sequences have been reported for PAPs. Binding efficiency was protein concentration dependent, but it was independent of the presence of other cleavage and polyadenylation factors (figure 1B). In human, PAPγ was not able to extent poly(A) tails on its own but required the interaction with CPSF [50]. Therefore, our results suggested that EhPAP binding and catalytic activities may be regulated through different mechanisms.

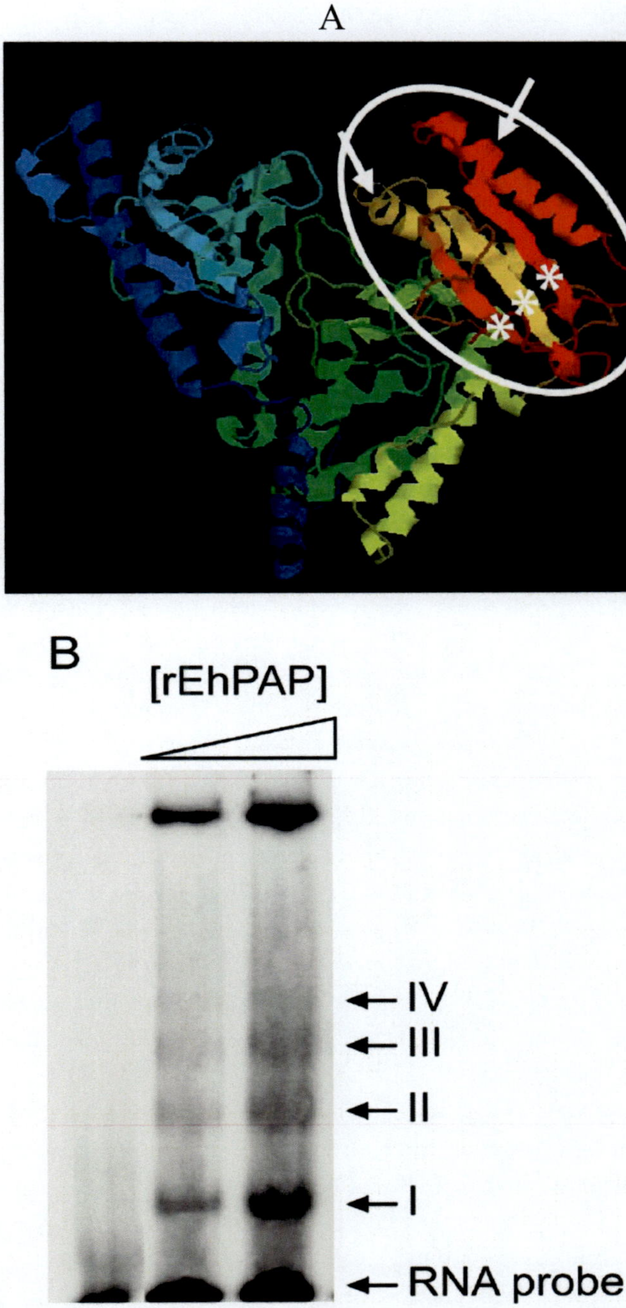

Figure 1. RNA binding activity of rEhPAP. A, EhPAP three dimensional structure. The prediction was created by Geno3D and Rasmol programs using bovine PAP PDB file as template. Blue, central domain; green, catalytic domain that contains the three invariant Asparte residues involved in nucleotide transfer and the F/YGS responsible for ATP binding; red, RNA-binding domain at the C-end (circled). Arrow, α-helix; star, β sheet. B, EhPAP RNA-binding activity. The purified rEhPAP was incubated with the [α-^{32}P] UTP-labeled PSIII[156] probe and complexes were resolved by 6% PAGE. Lane 1, free probe; lane 2, rEhPAP (50 ng); lane 3, rEhPAP (100 ng). I to IV, RNA-rEhPAP complexes.

4.2. Non Canonical Poly(A) Polymerases of *E. histolytica*

In order to gain insights into the E. histolytica poly(A) polymerases family, we searched for non canonical PAPs in the parasite genome. Putative orthologous proteins were selected from BLASTP analysis using the Blosum 62 scoring matrix (http://us.expasy.org/tools/blast/). Extensive in silico analysis led to the detection of six proteins that present 21-30% identity and 44-52% homology with H. sapiens and S. pombe unconventional PAPs, including GLD-2, Cid11, Cid13, Cid14 and TRF-4 proteins, that all exhibit poly(A) polymerase activity in vitro (Table 1). E-values are lower than 7e-04, confirming that we identified orthologous proteins. Based on BLASTP analysis results, three proteins, named here as EhGLD-2a, EhGLD-2b and EhGLD-2c, were considered as GLD-2 homologues, and three proteins, named here as EhTRF4-a, EhTRF4-b and EhTRF4-c, were classified as TRF4 homologues. In agreement with previous reports about the small number of introns in E. histolytica [53], only two genes, EhGLD-2a and EhGLD-2c, have introns with the conserved E. histolytica acceptor and donor sites (Table 1).

E. histolytica non canonical poly(A) polymerases were compared with several eukaryotic nuclear PAPs (including EhPAP) using the ClustalW software (http://www.ebi.ac.uk/clustalw/). The obtained multiple alignment showed that proteins sequences are highly divergent (data not shown), confirming that they are not homologous proteins. We next analyzed amino acids sequences using ScanProsite (http://us.expasy.org/tools/scanprosite/), InterProScan (http://www.ebi.ac.uk/InterProScan/) and Pfam (http://www.sanger.ac.uk/Software/Pfam/search.shtml) programs to predict structural domains and sequence patterns. As known for H. sapiens, S. cerevisiae, S. pombe and C. elegans non canonical PAPs, we identified in E. histolytica proteins a PAP domain with the conserved amino acids (F/YGS motif) involved in ATP binding, as well as the three aspartate residues responsible for the catalytic activity (figure 2A). In addition, non canonical PAPs do not have the RRM-like region described in nuclear PAPs. Finally, E. histolytica proteins are constituted by 287-456 aa, which is in the range of the molecular size reported for eukaryotic unconventional PAPs (figure 2B). Taken altogether, these data strongly suggested that EhGLD-2a, EhGLD-2b, EhGLD-2c, EhTRF4-a, EhTRF4-b and EhTRF4-c are truly non canonical poly(A) polymerases orthologues. It is possible that they require the interaction with another polypeptide carrying the RNA-binding domain to be functional, as demonstrated for C. elegans GLD-2 [13].

We next performed a phylogenetic study in order to determine the relationships between orthologous proteins through evolutionary scale. Amino acids sequences were aligned by the ClustalW software and the phylogenetic tree were constructed using the Neighbor-Joining distance method [54] as implemented in the MEGA package version 2.1 [55]. The robustness of the phylogenetic inferences was tested by bootstrapping method, involving 500 replications of the data based on the criteria of 50% majority-rule consensus. As represented in figure 2C, orthologous proteins are closely related but they form two separate clades, which both contain enzymes with reported in vitro polyadenylation activity. Group I corresponds to GLD-2 orthologues, whereas group II includes TRF4 orthologues. EhTRF4-a, EhTRF4-b and EhTRF4-c are highly related to TRF4 proteins, in agreement with homology values obtained from BLAST analysis (table 1). Although EhGLD-2a and

Table 1. Comparisons of *E. histolytica*, *H. sapiens* and *S. pombe* non canonical PAPs

E. histolytica protein		H. sapiens homologue					S. pombe homologue				
Name	Accession Number[a]	Name	Accession Number[b]	E-value	I	H	Name	Accession Number[b]	E-value	I	H
EhGLD-2a*	XM_650512	GLD-2	Q86WZ2	2e-08	21	41	CID13	O74518	3e-15	27	52
EhGLD-2b	XM_648269	GLD-2	Q86WZ2	4e-06	27	47	CID11	O13833	7e-04	27	46
EhGLD-2c*	XM_645947	GLD-2	Q86WZ2	3e-06	25	44	CID13	O74518	2e-15	30	51
EhTRF4-a	XM_643822	TRF4-2	Q8NDF8	1e-27	26	47	CID14	Q9UTN3	8e-20	27	46
EhTRF4-b	XM_643605	TRF4-2	Q8NDF8	1e-25	29	51	CID14	Q9UTN3	2e-24	28	52
EhTRF4-c	XM_643450	TRF4-2	Q8NDF8	2e-24	27	46	CID14	Q9UTN3	5e-28	27	48

* Intron containg gene. Identity (I) and Homology (H) values are expressed in percentage.
[a]GenBank database; [b]SwissProt/TrEMBL database.

EhGLD-2c were also clustered with TRF4 proteins, they are in a separate arm and seem to be in an intermediate position between GLD-2 and TRF4 proteins. Intriguingly, EhGLD-2b emerges from a distinct node, suggesting that this protein has evolved independently at a very early stage of the parasite evolution (figure 2C).

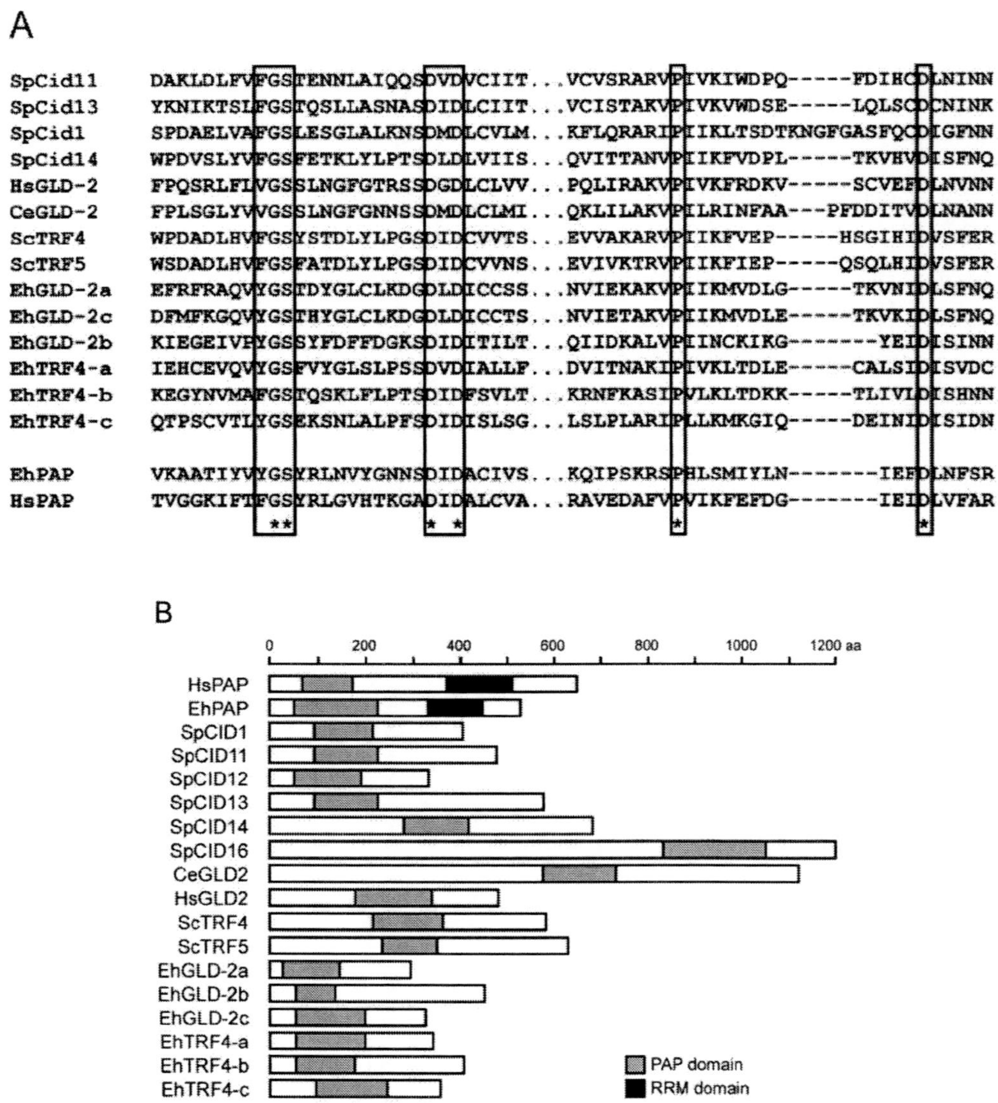

Figure 2. (Continued on next page)

C

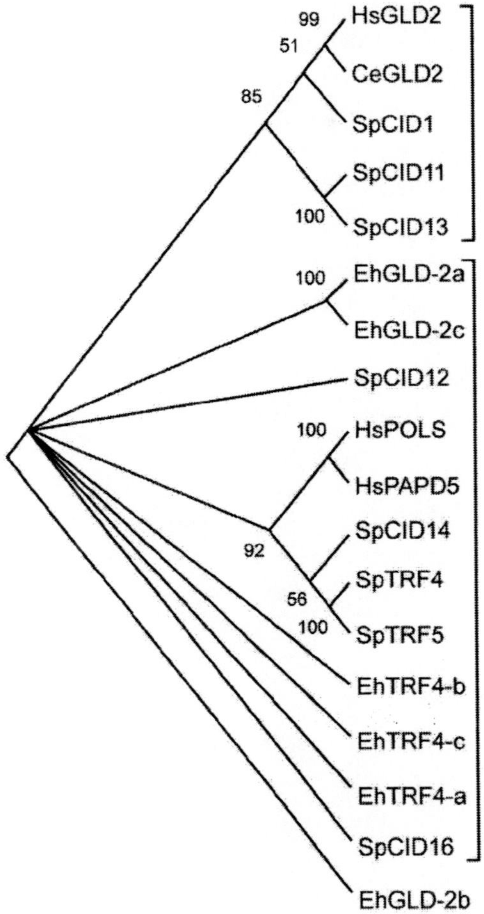

Figure 2. Comparisons of canonical and non canonical poly(A) polymerases. A, Multiple alignment of PAP domain. B, Schematic representation of PAP domains. C, Phylogenetic analysis.

Table 2. Oligonucleotides used in RT-PCR assays.

Gene	Sense oligonucleotide	Antisense oligonucleotide	Amplified product
EhGLD-2a	5′-GTAATGGATAGGTGGCTTCGG	5′-CAGCAGGAATAATGAATAACTCATC	185 bp
EhLD-2b	5′-ATGGAAGAAGAAGAACAATTTAATG	5′-TTTAAATAAATTAGGTTTGTCTGTTG	297 bp
EhGLD-2c	5′-ATGTCGATTTCAAACGATGG	5′-CTTGTCCTTTAAACATAAAATCAAC	190 bp
EhTRF4-a	5′-ATGGCAGCATGGTTAAATGG	5′-GGTAATTGAGGAAATAATAATGCAA	251 bp
EhTRF4-b	5′-CTTTGAAGGAGAATTAGATCTAAATC	5′-GCTTTAAAATTTCTTTTTTGGTC	297 bp
EhTRF4-c	5′-CACCTCAAATCAGTGATTATGC	5′-CTTACTTTTGGTTTTCCATTCTTA	254 bp

To verify whether these non canonical PAPs were expressed in E. histolytica trophozoites, we analyzed mRNA expression through RT-PCR assays. Trophozoites (clone A) were axenically cultured in TYI-S-33 medium [56] and total RNA was obtained using TRIzol™ (Gibco) reagent. Semi quantitative RT-PCR was performed as described [45] with forward and reverse specific oligonucleotides that amplify an internal region of each gene (table 2). Actin primers were used as an internal control in the same experiments. Amplified products were separated by 10% PAGE, stained with ethidium bromide and visualized through a UV-transiluminator. As shown in figure 3A, we were able to obtain amplified products of the expected size for the six E. histolytica non canonical Pap genes. Interestingly, mRNA expression varies among this gene family: EhTRF4-a is highly expressed; EhTRF4-c, EhGLD-2a and EhGLD-2c have a moderate expression; whereas EhTRF4-b and EhGLD-2b are poorly expressed. Densitometric analysis and normalization against actin product confirmed that EhTRF4-a expression is about 1.75-fold actin expression; EhGLD-2a, EhGLD-2c and EhTRF4-c expression correspond to 0.8 to 1.2-fold actin expression; whereas EhGLD-2b and EhTRF4-b expression is only 0.2 and 0.1-fold actin expression, respectively (figure 3B). The differential expression of E. histolytica non canonical PAPs suggested that they may participate in distinct cellular processes in trophozoites. It should be interesting to evaluate their expression through cell cycle progression in order to determine if they participate in this event as it has been described in the unicellular organism S. pombe.

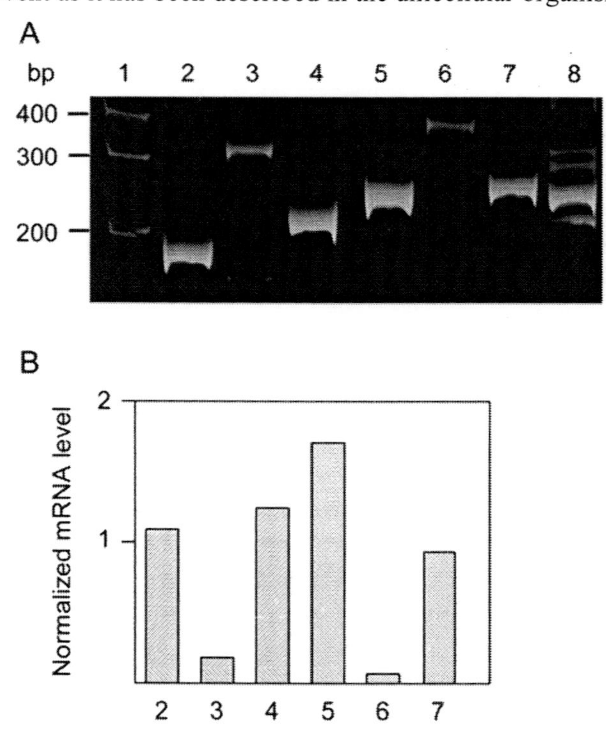

Figure 3. mRNA expression of non canonical *E. histolytica* PAPs. A, Semi-quantitative RT-PCR assays using specific internal primers and total RNA from *clone A trophozoites. Lane 1, molecular markers; lane 2, EhGLD-2a; lane 3, EhGLD-2b*; lane 4, *EhGLD-2c;* lane 5, *EhTRF4-a*; lane 6, *EhTRF4-b*; lane 7, *EhTRF4-c*; lane 8, *actin*. B, Densitometric analysis of bands in A. Pixels corresponding to *actin* product were taken as 100% and *Paps genes* products levels were expressed with respect to *actin* amount. Numbers are as in A.

4.3. Relationships between Canonical and Non Canonical PAPs

Poly(A) polymerases and other enzymes with distinct biochemical properties (DNA polymerases-β, CCA-adding enzymes, 2′-5′ oligo(A) synthetases, protein nucleotidyltransferases and terminal deoxynucleotidyltransferases) belong to the polymerase β-like superfamily, the most widespread superfamily of nucleotidyltransferases that all catalyze the transfer of a nucleotide to a acceptor hydroxyl group. This is a key reaction in many fundamental biological processes including DNA replication and repair, transcription and RNA processing, as well as other specialized pathways. Polymerase β-like nucleotidyltransferases do not share high similarity, but they were initially characterized by the following amino acid residue pattern, hGGSx(9-13)DhDEh (where h indicates an hydrophobic amino acid and x is any amino acid) [57].

In order to initiate the characterization of the polymerase β-like superfamily of nucleotidyltransferases in E. histolytica, we studied the structural and evolutionary relationships between canonical and non canonical PAPs through PSI-BLAST analysis. This program generates a weighted profile from the sequences detected in the first pass of a gapped-BLAST search and iteratively searches the database using this profile as the query, allowing the inclusion of sequences with E-value cut off higher than 0.01 [58]. Canonical and non canonical E. histolytica poly(A) polymerases were submitted to PSI-BLAST analysis to identify related polyadenylases in E. histolytica, mammals and bacteria. Using the E-value threshold as a as a similarity measure, we evidenced a clear relation between E. histolytica unconventional PAPs (figure 4). For example, EhTRF4-a is highly related to EhTRF4-b, EhTRF4-c, EhGLD-2a and EhGLD-2c, but it presents less similarity with EhGLD-2b. This is in agreement with the phylogenetic tree where EhGLD-2b is in a separated arm, indicating that this protein resulted from a distinct evolution event. As expected, E. histolytica unconventional PAPs do not have relationships with the nuclear EhPAP, which is highly related to other eukaryotic canonical proteins. The bacterial PAP remains isolated. Interestingly, we showed that E. histolytica non canonical and canonical PAPs, as well as bacteria and other eukaryotic canonical PAPs, are all related to the minimal nucleotidyltransferase (MNT) domain. This conserved functional domain was initially identified in eukaryotic members of the polymerase β-like superfamily of nucleotidyltransferases, through iterative database searches, multiple alignments of primary sequences and secondary structure predictions. Then, it was detected in bacterial polymerases, suggesting that the evolution of the polymerase β-like superfamily of nucleotidyltransferases may have included bursts of rapid divergence linked with the emergence of new functions as well as a number of horizontal gene transfer events [57]. The presence of this ancestral bacterial MNT domain in E. histolytica canonical and non canonical PAPs confirmed that these proteins belong to the polymerases β-like superfamily of nucleotidyltransferases. Their functional characterization will help us to understand their biological role in E. histolytica gene expression regulation.

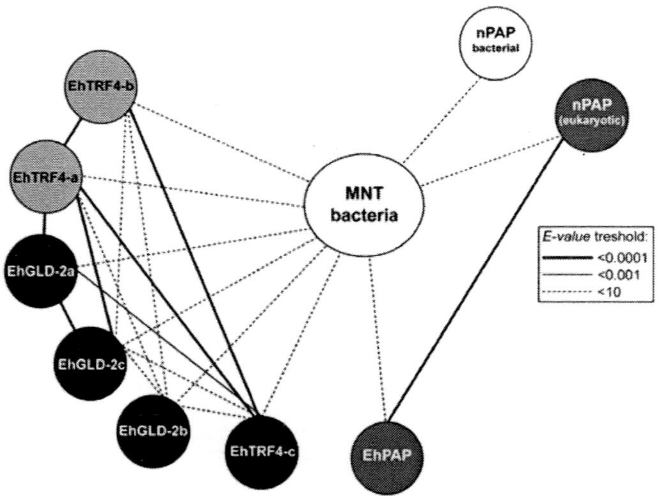

Figure 4. The *E. histolytica* poly(A) polymerases family. Proteins relationships were evaluated through PSI-BLAST analysis. The width of connecting lines indicates similarity level.

5. Conclusion

The protozoan parasite Entamoeba histolytica has a large poly(A) adding enzymes family that is constituted by one canonical poly(A) polymerase and six non canonical poly(A) polymerases, indicating that mRNA poly(A) addition is a relevant regulatory event that is conserved through evolutionary scale. Although they do not share high amino acids similarity, E. histolytica non canonical PAPs have the conserved hallmarks of the DNA polymerases β-like superfamily of nucleotidyltransferases, including the specific amino acids involved in nucleotide interaction and catalysis. The diversity of poly(A) polymerases in E. histolytica suggests that mRNA poly(A) tail synthesis and mRNA turnover are important control points for translation efficiency in this organism. The further characterization of canonical and non canonical enzymes would provide new insights into the molecular mechanisms involved in gene expression regulation in this protozoan parasite.

Acknowledgements

Preliminary sequences were obtained from TIGR, Sanger and Pathema E. histolytica genome databases. The E. histolytica sequencing effort is part of the Genome Sequencing Project and is supported by awards from The National Institute of Allergy and Infectious Diseases, National Institute of Health, and The Welcome Trust. This work was supported by UACM (México), CONACyT (México), COFAA-IPN (Mexico), SIP-IPN (Mexico) and European Community grants. We thank Alfredo Padilla for his help in the artwork.

REFERENCES

[1] Colgan, D; Manley, J. Mechanism and regulation of mRNA polyadenylation. *Genes and Development*, 1997 11, 2755-2766.

[2] Huang, Y; Carmichael, G. Role of polyadenylation in nucleocytoplasmic transport of mRNA. *Molecular and Cellular Biology*, 1996 16, 1534–1542.

[3] Tarun, SZ Jr; Sachs, AB. Association of the yest poly(A) tail binding protein with translation initiation factor eIF-4G. *EMBO Journal*, 1996 15, 7168–7177.

[4] Tarun, SZ Jr; Wells, SE; Deardorff, JA; Sachs, AB. Translation initiation factor eIF4G mediates *in vitro* poly(A) tail-dependent translation. *The Proceedings of the National Academy of Sciences,* 1997 94, 9046–9051.

[5] Tucker, M; Valencia-Sánchez, MA; Staples, RR; Chen, J; Denis, CL; Parker, R. The transcription factor associated Ccr4 and Caf1 proteins are components of the major cytoplasmic mRNA deadenylase in *Saccharomyces cerevisiae*. *Cell,* 2001 104, 377-386.

[6] Keller, W; Bienroth, S; Lang, KM; Christofori, G. Cleavage and polyadenylation factor CPF specifically interacts with the pre-mRNA 3' processing signal AAUAAA. *EMBO Journal,* 1991 10, 4241-4149.

[7] Kaufmann, I; Martin, G; Friedlein, A; Langen, H; Keller, W. Human Fip1 is a subunit of CPSF that binds to U-rich RNA elements and stimulates poly(A) polymerase. *EMBO Journal,* 2004 23, 616-626.

[8] Kerwitz, Y; Kuhn, U; Lilie, H; Knoth, A; Scheuermann, T; Friedrich, H; Schwarz, E; Wahle, E. Stimulation of poly(A) polymerase through a direct interaction with the nuclear poly(A) binding protein allosterically regulated by RNA. *EMBO Journal,* 2003 22, 3705-3147.

[9] Proudfoot, NJ; Furger, A; Dye, MJ. Integrating mRNA processing with transcription. *Cell,* 2002 108, 501-512.

[10] Colgan, DF; Murthy, KG; Prives, C; Manley, JL; Cell-cycle related regulation of poly(A) polymerase by phosphorylation. *Nature,* 1996 384, 282-285.

[11] Zhao, W; Manley, JL. Desregulation of poly(A) polymerase interferes with cell growth. *Molecular and Cellular Biology*, 1998 18, 5010-5020.

[12] Martin, G; Keller, W; Doublie, S. Crystal structure of mammalian poly(A) polymerase in complex with an analog of ATP. *EMBO Journal*, 2000 19, 4193-4203.

[13] Wang, L; Eckmann, CR; Kadyk, LC; Wickens, M; Kimble, J. A regulatory cytoplasmic poly(A) polymerase in *Caenorhabditis elegans*. *Nature*, 2002 419, 312–316.

[14] Kwak, JE; Wang, L; Ballantyne, S; Kimble, J; Wickens, M. Mammalian GLD-2 homologs are poly(A) polymerases. *The Proceedings of the National Academy of Sciences,* 2004 101, 4407–4412.

[15] Eckmann, C; Kraemer, B; Wickens, M; Kimble, J. GLD-3, a Bicaudal-C homolog that represses FBF to control germline sex determination in *C. elegans*. *Developmental Cell,* 2002 3, 697-710.

[16] Barnard, DC; Ryan, K; Manley, JL; Richter, JD. Symplekin and xGLD-2 are required for CPEB-mediated cytoplasmic polyadenylation. *Cell*, 2004 119, 641–651.

[17] Rouhana, A; Wang, L; Buter, N; Kwak, JE; Schiltz, CA; Gonzalez, T; Kelley, AE; Landry, CF; Wickens, M. Vertebrate GLD-2 poly(A) polymerases in the germline and the brain, *RNA*, 2005 11, 1117–1130.

[18] Read, RL; Martinho, RG; Wang, SW; Carr, AM; Norbury, CJ. Cytoplasmic poly(A) polymerases mediate cellular responses to S phase arrest. *The Proceedings of the National Academy of Sciences,* 2002 99, 12079-12084.

[19] Castano, IB; Heath-Pagliuso, S; Sadoff, BU; Fitzhugh, DJ; Christman, MF. A novel family of TRF (DNA topoisomerase I-related function) genes required for proper nuclear segregation. *Nucleic Acids Research*, 1996 24, 2404-2410.

[20] Sadoff, BU; Heath-Pagliuso, S; Castaño, IB; Yingfang, Z; Kieff, FS; Christman, MF. Isolation of mutants of *Saccharomyces cerevisiae* requiring DNA topoisomerase I. *Genetics*, 1995 141, 465–479.

[21] Huh, WK; Falvo, JV; Gerke, LC; Carroll, AS; Howson, RW; Weissman, JS; O'Shea, EK. Global analysis of protein localization in budding yeast. *Nature*, 2003 425, 686–691.

[22] Richter, JD. Cytoplasmic polyadenylation in development and beyond *Microbiology and Molecular Biology Reviews*, 1999 63, 446-456.

[23] Mendez, R; Murthy, KG; Ryan, K; Manley, JL; Richter, JD. Phosphorylation of CPEB by Eg2 mediates the recruitment of CPSF into an active cytoplasmic polyadenylation complex. *Molecular Cell*, 2000 6, 1253–1259.

[24] Rouget, C; Papin, C; Mandart, E. Cytoplasmic CstF-77 protein belongs to a masking complex with CPEB in Xenopus oocytes. *Journal of Biological Chemistry,* 2006, in press.

[25] Stebbins-Boaz, B; Richter, JD. Translational control of the cleavage and polyadenylation specificity factor in *Xenopus* during early development. Critical *Reviews in Eukaryotic Gene Expression* 1997 *7*, 73–94.

[26] Dickson, KS; Bilger, A; Ballantyne, S; Wickens, MP. The cleavage and polyadenylation specificity factor in *Xenopus laevis* oocytes is a cytoplasmic factor involved in regulated polyadenylation. *Molecular and Cellular Biology*, 1999 *19*, 5707–5717.

[27] Huarte, J; Stutz, A; O'Connell, ML; Gubler, P; Belin, D; Darrow, AL; Strickland, S; Vassalli, JD. Transient translational silencing by reversible mRNA deadenylation. *Cell,* 1992 69, 1021-1030.

[28] Hake, LE; Richter, JD. Translational regulation of maternal mRNA. *Biochimica et Biophysica Acta,* 1997 1332, M31-M38.

[29] Wu, L; Wells, D; Tay, J; Mendis, D; Abbott, MA; Barnitt, A; Quinlan, E; Heynen, A; Fallon, JR; Richter, JD. CPEB-mediated cytoplasmic polyadenylation and the regulation of experience-dependent translation of a-CaMKII mRNA at synapses. *Neuron*, 1998 21, 1129–1139.

[30] Alarcon, JM; Hodgman, R; Theis, M; Huang, YS; Kandel, ER; Richter, JD. Selective modulation of some forms of schaffer collateral-CA1 synaptic plasticity in mice with a disruption of the CPEB-1 gene. *Learning and Memory*, 2004 11, 318–327.

[31] Saitoh, S; Chabes, A; McDonald, WH; Thelander, L; Yates, JR; Russell, P. Cid13 is a cytoplasmic poly(A) polymerase that regulates ribonucleotide reductase mRNA. *Cell*, 2002 109, 563-573.

[32] Wang, SW; Toda, T; MacCallum, R; Harris, AL; Norbury, C. Cid1, a fission yeast protein required for S-M checkpoint control when DNA polymerase delta or epsilon is inactivated. *Molecular and Cellular Biology*, 2000 20, 3234–3244.

[33] Wang, Z; Castano, IB; De Las Penas, A; Adams, C; Christman. M.F. Pol kappa: a DNA polymerase required for sister chromatid cohesion. *Science*, 2000 289, 774–779.

[34] Zhenghe Wang, Z; Castano, IB; Adams, C; Vu, C; Fitzhugh, F; Christman, MF. Structure/Function Analysis of the *Saccharomyces cerevisiae* Trf4/Pol DNA Polymerase. Genetics, 2002 160, 381–391.

[35] Haracska, L; Johnson, RE; Prakash, L; Parcash, S. Trf4 and Trf5 proteins of *Saccharomyces cerevisiae* exhibit Poly(A) RNA polymerase activity but no DNA polymerase activity. *Molecular and Cellular Biology*, 2005, 25 10183-10189.

[36] van Hoof, A; Lennertz, P; Parker, R. Yeast exosome mutants accumulate 3′-extended polyadenylated forms of U4 small nuclear RNA and small nucleolar RNAs. *Molecular and Cellular Biology,* 2000 20, 441–452.

[37] Kuai, L; Fang, F; Butler, JS; Sherman, F. Polyadenylation of rRNA in *Saccharomyces cerevisiae. The Proceedings of the National Academy of Sciences*, 2004 101, 8581–8586.

[38] Kadaba, S; Krueger, A; Trice, T; Krecic, AM; Hinnebusch, AG. et al. Nuclear surveillance and degradation of hypomodified initiator tRNAMet in S. cerevisiae. *Genes and Development*, 2004 18, 1227–1240.

[39] Vanacova, S; Wolf, J; Martin, G; Dettwiller, S; Friedlein, A; Langen, H; Keith, G; Keller, W. A new yeast poly(A) polymerase complex involved in RNA quality control. *PLOS Biology*, 2005 3, 986-997.

[40] Houseley, J; Tollervey, D. Yeast Trf5p is a nuclear poly(A) polymerase. *EMBO reports,* 2006 7, 205-211.

[41] Jackson, TF. (2000). Epidemiology In Amebiasis. Ravdin, J.I. (ed.). London: Imperial College Press, 47-63.

[42] Gilchrist, C; Megan, L; Genghis, L; Mann. B; Petri Jr W. Calcium Modulates Promoter Occupancy by the *Entamoeba histolytica* Ca^{2+}-binding Transcription Factor URE3-BP. *Journal of Biological Chemistry,* 2003 278, 4646-4653.

[43] Marchat, LA; Gomez, C; Perez, D; Paz, F; Mendoza, L; Orozco E. Two CCAAT/enhancer binding protein sites are cis-activator elements of the Entam*oeba histolytica EhPgp1* (*mdr*-like) gene expression. *Cellular Microbiology,* 2002 4, 725-737.

[44] de Dios-Bravo, G; Luna-Arias, JP; Riveron, AM; Olivares-Trejo, JJ; Lopez-Camarillo, C; Orozco, E. *Entamoeba histolytica* TATA-box binding protein binds to different TATA variants *in vitro*. FEBS Journal, 2005 272, 1354-1366

[45] Lopez-Camarillo, C; Luna-Arias, JP; Marchat, LA; Orozco, E. *EhPgp5* mRNA stability is a regulatory event in the *Entamoeba histolytica* multidrug resistance phenotype. *Journal of Biological Chemistry*, 2003 278, 11273-11280.

[46] Loftus, B; Anderson, I; Davies, R; Alsmark, UC; Samuelson, J. et al. The genome of the protist parasite *Entamoeba histolytica. Nature*, 2005 433, 865-868.

[47] Lopez-Camarillo, C; Orozco, E; Marchat, LA. *Entamoeba histolytica*: comparative genomics of the pre-mRNA 3' end processing machinery, *Experimental Parasitology*, 2005 110, 184-190.

[48] Bruchhaus, M; Leippe, C; Lioutas, C; Tannich, E. Inusual gene organization in the protozoan parasite *Entamoeba histolytica. DNA and Cell Biology*, 1993 12, 925–933.

[49] Garcia-Vivas, J; Lopez-Camarillo, C; Azuara-Liceaga, E; Orozco, E;. Marchat, LA. *Entamoeba histolytica*: cloning and expression of the poly(A) polymerase EhPAP, *Experimental Parasitology,* 2005 110, 226-232.

[50] Kyriakopoulou, CB; Nordvarg, H; Virtanen, A. A novel nuclear human poly(A) polymerase (PAP), PAP . *Journal of Biological Chemistry*, 2001 276, 33504-33511.

[51] Morel, AP; Sentis, S; Bianchin, C; Le Romancer, M; Jonard, L; Rostan, MC; Rimokh, R; Corbo, L. BTG2 antiproliferative protein interacts with the human CCR4 complex existing in vivo in three cell-cycle-regulated forms. *Journal of Cell Science,* 2003 116, 2929-3296.

[52] Schul, W; Driel, RV; Jong, LD. A subset of poly(A) polymerase is concentrated at sites of RNA synthesis and is associated with domains enriched in splicing factors and poly(A) RNA. *Experimental Cell Research*, 1998 238, 1-12.

[53] Willhoeft, U; Campos-Gongora, E; Touzni, S; Bruchhaus, I; Tannich, E. Introns of *Entamoeba histolytica* and *Entamoeba dispar. Protist,* 2001 152, 149-156.

[54] Saitou, N; Nei, M. The neighbor-joining method: a new method for reconstructing phylogenetic trees. *Molecular Biology and Evolution,* 1987 4, 406-425.

[55] Kumar, S; Tamura, K; Nei, M. MEGA3: Integrated Software for Molecular Evolutionary Genetics Analysis and Sequence Alignment. *Briefings in Bioinformatics,* 2004 5, 150-163.

[56] Diamond, LS; Harlow, R; Cunnick, C. A new medium for the axenic cultivation of *Entamoeba histolytica* and another *Entamoeba*. Transactions of the Royal Society of Tropical Medicine and Hygiene*,* 1978 73, 431-432.

[57] Aravind, L; Koonin, EV. DNA polymerase β-like nucleotidyltransferase superfamily: identification of three new families, classification and evolutionary history. *Nucleic Acids Research,* 1999 27, 1609–1618.

[58] Alschult, SF; Madden, TL; Schaffer, AA; Zang, J; Zang, Z; Miller, W; Lipman, DJ. Gapped BLAST and PSI-BLAST: a new generation of protein database search programs. *Nucleic Acid Research*, 1997 25, 3389-3402.

Index

A

B

C

D

J

K

L

M

N

O

P

T